Science and the Unborn

SCIENCE
AND
THE UNBORN

Choosing Human Futures

CLIFFORD GROBSTEIN

Basic Books, Inc., Publishers

NEW YORK

Library of Congress Cataloging-in-Publication Data

Grobstein, Clifford, 1916–
    Science and the unborn / Clifford Grobstein.
        p.   cm.

    Includes index.
    ISBN 0-465-07295-X : $18.95
    1. Fetus—Research—Moral and ethical aspects.   2. Embryology,
Human—Research—Moral and ethical aspects.   3. Unborn children
(Law) I. Title.
RG600.G76   1988
174′.2—dc19   88-47667
                                                                            CIP

CONTENTS

PREFACE

IN THIS BOOK I address the status of the unborn, a matter that has been, in the past decade or two, highly contentious with respect to abortion. By declaring the unborn to be persons from the moment of conception, antiabortion forces have sought to outlaw the procedure, in hopes of protecting unborn lives against termination by maternal choice. Their opponents, with equal vigor, have insisted on the right of women to control their reproduction, without challenge by opposing rights of the unborn.

In 1973 the United States Supreme Court, in *Roe v. Wade*, a landmark case, denied the antiabortion position on constitutional grounds. While staking out a firm legal foundation, the Court's decision did little to relieve the political battle. Instead, the decision led to trench warfare that continues today.

Meanwhile, other developments in reproductive biology and medicine have raised new questions about the status of the unborn. It has become increasingly clear that public policy in this area needs firmer definition if continuing controversy is not to inhibit scientific and medical advances in the public interest.

I chose to write about science and the unborn because I am a biological scientist with a background of research in mammalian embryology who has had considerable exposure to medical affairs. During the past decade my professional activity has focused on the impacts of advancing biomedical science on such public policy questions as regulation of genetic technology and of external human fertilization and preembryo transfer (*in vitro* fertilization, or IVF). I have come to believe that in this general area, no matter needs more scientific clarification and greater public understanding than the nature of the unborn.

Of course, scientific facts alone cannot dictate resolution of such emotionally wrenching issues as abortion—any more than can a decision of the Supreme Court. But abortion is no longer the only issue on which the moral, legal, and social status of the unborn weighs heavily. Moreover, further issues can be expected to arise in the not too distant future, driven by the dramatic advance of biomedical science during the second half of this scientifically enormously productive century. The advances are exciting and satisfying to those who look to knowledge as a source of human progress. But new knowledge also brings new questions and new risks, as this century is also demonstrating. Among other things, new knowledge is bringing us ever more frequently into face-to-face confrontation with the previously almost entirely covert unborn.

How should we regard and treat the unborn? In other words, what should their status be? Should they be treated like the rest of us—in provision of medical care, in human rights, in level of social investment? Or should they remain in an ambiguous policy limbo, as if waiting in the wings (actually the womb) for their appointed entrance but with no role in the already ongoing play?

Their ambiguity is reflected by the fact that they have no name and one does not usually know whether to refer to her, him, or it. However they may be called or referred to, almost everyone tends to be vague as to any detail of their state at any given time. From a fertilized egg (a single cell) to the enormous complexity of the late fetus, they lie hidden in the womb (or are part of an anonymous but perhaps embarrassingly obvious prenatal bulge). "Unborn" is a thoroughly negative, noncommunicative, black-box kind of term. It does not say a thing about what *is*, only about what *is not*. The remarkable, almost fantastic, transformation of a single cell into a newborn babe—the whole seeming miracle of becoming—is concealed and obscured in the cloistered and segregated existence of the unborn.

That is almost more than an embryologist can bear. It is like hiding the Mona Lisa under a shroud. In this book I want to dispel the darkness of the womb to illuminate what is within— not because that will resolve the abortion dilemma but rather to reveal the true nature of the process of becoming that unfolds unseen. In turn, a clearer view may offer a clearer definition of how to treat the unborn when decisions about them are required.

However, this book does not describe all there is to know about human embryology. To do so would require a far more detailed and much longer work. Rather, I attempt selectively to review what we know about human development (often derived from analogy with other animals) and to apply this knowledge to the question of the status of the unborn in terms of public policy. I emphasize that during the unborn period the individuality that we so prize in our society, and that is so central to the concept of a person, is progressively emerging. In truth, just as we can trace the genesis of the heart, we can trace the genesis of individuality. The matter is more difficult with personal individuality because the phenomenon and the concept are more complex.

The full course of development encompasses at least six reasonably different aspects: genetic, developmental, functional, behavioral, psychic, and social. Each must be distinguished and considered separately and then together, in order to understand the whole process of which they are part. We shall see that they are all involved in the controversy over status, with different perspectives (the developing entity, the mother, the scientist, the judge) making one or another of the aspects, or their combination, especially significant to status.

As an embryologist, I cannot imagine that appropriate assignment of moral and social status can ignore the nature and developmental course of individuality. On the other hand, our purposes and objectives in assigning status are at least equally fundamental. The question of human status does not arise as an abstraction, it arises in specific contexts where the question is: How may I treat these embryonic cells, this embryo, this fetus? Whose interests must be regarded, whose is the responsibility? More specifically, should externally fertilized eggs be discarded when they can no longer be transferred to the donor's uterus? Should corrective surgery be attempted on a malformed fetus in the uterus, even if there is significant risk to the welfare or life of the mother?

It is to be emphasized again that scientific facts alone cannot answer such questions. They may arise in a laboratory or clinic, but they have another dimension in the domains of morality and social purpose. But abstract moral principle or bloodless constitutional dogma are no better by themselves. In practice, these and all other relevant considerations must be applied to particular people in particular situations and in particular states of being. The matter ends up as a multiplex human judgment, with many factors having their role.

The general policy question thus becomes one of the nature of the decision process, especially identification of the roles of all participants. Yet the decision process and its result must not offend the overall public ethos. That is the demanding requirement properly imposed by a moral society, no matter how pluralistic.

As previously noted, it is the recent advance of reproductive science and technology that has markedly raised consciousness about the unborn and their treatment. This advance has put physicians, scientists, and technologists in a conspicuous role in relation to decision making. Yet their role cannot be that of final arbiters in reaching decisions. The questions involved extend far beyond particular expertise or any single perspective. What is needed is a widely accessible public policy process to define the shape of the permissible, even reflecting significant societal uncertainty where it exists. Without this, the connectedness of ethos and public action may be lost, leaving public policy without moral foundation.

Given its prominence as an agent of change, the special role and contribution of science is to make generally available the most reliable relevant knowledge. Even though such knowledge alone is not sufficient to resolve heavily value-laden issues, it can at least provide a commonly shared foundation. Even if full agreement is not achieved, better-informed perspectives may offer openings for accommodation and consensus. At the least, policy based on such information is less likely to be challenged for nonconformity with established knowledge.

It is vital that we attempt such an approach at the present time. Differences over abortion have aroused conflicts so deep that the effects spread and inflame many other areas of public policy. Almost unnoticed in the din, however, the battleground is shifting under the entrenched contestants over abortion. New technological innovations have brought the unborn into more general view and in contexts other than the termination of pregnancy. The importance of establishing a wider and sharper perception of the unborn period is thereby increased.

For example, IVF has laid bare in a laboratory dish the primordial step in initiating a new individual—the union of sperm and egg in fertilization. The first few divisions of this primordial zygote cell are routinely available for examination and are accessible to manipulation. The subsequent transfer of such early entities to the maternal uterus has enabled several thousand otherwise barren couples around the world to produce a child; a small number of these transfers have occurred after frozen storage and thawing of preembryos.

This new accessibility of the earliest stages of human development, together with the ability to transfer them into the uterus, raises the question of status in a very different context from that of abortion. The purpose and result of IVF is not to terminate life but to continue it. It is not a procedure of mixed benefit and harm, but one that brings unalloyed joy to the couple involved.

Abortion aside, therefore, we must now ask how we should relate to these earliest stages in the human life history as we confront them for the first time outside the human body and on the very threshold of becoming. Who is responsible for them if they cannot be returned to the maternal uterus? Who is empowered or required to make decisions about them if their biological parents cannot or will not? Are they already individuals in the sense of human rights? What does their potential to become individuals, in the human rights sense, mean, and to what does it entitle them?

IVF and other recent advances have raised these tough issues anew. Current U.S. policy on the status of the unborn provides no definitive answers. The questions require that the nature of the unborn, as currently understood, be specified with greater clarity and force. How can one do this when not only the questions but the substantive matters involved are so fundamental, complex, and often deeply contentious?

The book proceeds as follows: It first examines the significance of status, viewing it as a thumbnail designation of how we should treat individual members of our society. The discussion highlights the special status problem presented by the unborn— that they are rapidly passing through fundamental transformations that are not fully understood but involve a *progression of states* that increasingly mobilizes interest and concern about individuals as persons.

The next step outlines what we know about four major phases of human development: preembryo, embryo, fetus, and neonate. Emphasis is placed on the major characteristics of each phase with reference to the kind of treatment each appears to warrant. This requires confronting not only the immediate issues raised but the wider and longer-range objectives that are the source of moral tension and conflict. At stake, therefore, is the political feasibility of specific policy options in the light of our heterogeneous collective ethos.

Finally, I attempt an overall preliminary approach to policy formation. My approach relies heavily on stepwise assignment of status during the developmental progression of the unborn as well as on a policy process that permits continuing evaluation and modification as new issues emerge. Such an approach takes into account both the gradual elaboration of fundamental human characteristics during development and the differing degrees of urgency in the policy issues that can be expected to come into future focus.

The rationale assumes, first, that development generates new properties central to status and that, therefore, its definition would best proceed stepwise prenatally as it does postnatally. Second, it assumes that our knowledge of critical characteristics that are fundamental to status—for example, sentience—is still incomplete but growing. Continuing rather than definitive decision processes are, accordingly, in order. Especially required is greater understanding of the functional maturation of the brain in relation to the emergence of inner subjective life, a subject that is inherently difficult.

One additional general observation is worth making about the general thrust of the book. The last several decades have witnessed a steady rise in the number and importance of major policy issues that are closely tied to scientific and technological advances. Examples appear in virtually all aspects of such public policy concerns as health, environmental protection, food safety, and military security. In a society that is increasingly technologically oriented, public policy cannot be shaped effectively without judicious fusion of the perspectives derived from science and technology with those of the general culture, particularly its value structure. In this sense, science and technology must be fully interactive with values, aspirations, and purposes. Nowhere, certainly, is this more the case than in considering the status of emerging members of the human family. Here it is especially clear that scientific fact is value-laden and morality must be science-aware. Though morality and science may each —for some purposes—have its own distinct primary realm, neither can be excluded in considering the course of a human future.

A number of people have made significant contributions to this book, some unknowingly. In the latter category are undergraduate students in several senior seminars who wrestled with the issues involved and found both frustration and satisfaction in achieving some consensus in a welter of initially diverse opinions. Observing and participating in this process deepened my sense of the human quality of the issues involved and the deep emotions they arouse.

Michael Flower, an associate along with John Mendeloff in a policy research project on reproductive options supported by the National Science Foundation (Grant PRA-8020679) from 1983 to 1986, traded ideas and facts with me over several productive years. His sensitivity to the ethical issues involved fine-tuned my own reactions and thinking substantially.

Similarly, jousting with fellow-members of the Ethics Committee of the American Fertility Society in preparing its report widened my perspective on legal, ethical, and medical professional attitudes that bear on the matter of unborn status.

And my wife, Ruth, not only kept me within the bounds of medical reality but tolerated regular desertion of our bed during the many nights devoted to preparation of the manuscript.

I am grateful to all of these contributors as well as to my many predecessors who have shared in print their struggles with these difficult issues.

CLIFFORD GROBSTEIN

Rancho Santa Fe, California
February 1988

Science and the Unborn

CHAPTER 1

The Significance
of Status

$\mathbf{F}$OR MOST PEOPLE, the unborn are an enigma. They both illustrate a general problem and have their own special difficulty. The general problem is: How does a new phenomenon arise when nothing resembling it seems to precede it, that is, when it seems to originate *de novo*?

An astrophysical example is the origin of a star out of interstellar dust. Astrophysicists have proposed on theoretical grounds that this happens, and astronomers are now observing what appear to be stages in the process. Dust falls together as the result of gravitational attraction, heating up as it condenses until the mass ignites to temperatures at which nuclear reactions begin. The result is a blazing star—for example, our sun. A star is not budded from an earlier star, it undergoes its own genesis from nonstar materials—some of which may be products of disintegration from earlier stars.

An anthroposocial instance is the initiation of a settled community at the intersection of two or more trade routes. When the numbers of travelers passing through the intersection reach some critical level, services provided to and paid for by the travelers can become the economic base for a population of nontravelers, the beginning of a community.

A developmental example is a seedling breaking out of its protective coats to become a tree. Without previous experience, who would guess that the tiny generative part of the seed is the progenitor of a massive sequoia?

In each case something entirely new seems to arise out of almost nothing—the source of the process does not even remotely resemble the end point. How do complex phenomena arise as if out of nothing? Is this evidence of divine guidance—a miracle without natural cause? Can it happen by chance? Or is the whole phenomenon already preformed at the beginning, in miniature, needing only to be scaled up in size to be visible?

Choice among these explanations has been debated for centuries. Today we are beginning to understand how the seeming miracle occurs. I shall say much more about that in later discussion. For the moment I note only that the transformations undergone by the human unborn provide this kind of enigma, and undoubtedly in the most complex and difficult form that we know.

In roughly nine calendar months a single human cell, the zygote produced at conception, or fertilization, multiplies in cell number, becomes cohesive, reorders, grows further, differentiates, and forms a multiplicity of functioning parts. Throughout this orderly and as-yet only incompletely comprehended process, the developing entity maintains its initial endowment of individuality and elaborates it into a human infant composed of billions of cells that are organized into one of the most intricate entities known to science. It is this emerging human entity, during its profound transformations, that must be given its due and appropriate place in the social fabric. An appropriate place means that it should have an accepted and commonly understood status, or site and role in its community. This is simple enough to declare; it proves to be no mean task to accomplish.

The Significance of Status

Status is the issue for the unborn because for them it is incompletely and poorly defined. Whether codified in law, usage, or tradition, status helps people to know how to treat and relate to each other. It may be as formal as military rank or as informal as the establishment of leadership by a strong personality on a playground. However established, there is no more fundamental status in moral, legal, and policy terms than membership in a human community. A formally defined status for the unborn is therefore essential to provide guidelines on how these nascent individuals should be treated throughout their developmental course from fertilized egg to birth.

The assignment of status to the unborn is especially difficult because not only are they an enigma, but they are, in effect, rapidly moving targets. They are undergoing fundamental transformations; they are continually changing and becoming something new. The pervasiveness, the magnitude, and yet the seamlessness of the changes during human development defy sharp and convenient classification. Moreover, this radical metamorphosis of properties and characteristics normally takes place while hidden in the seclusion of the womb. The process of creative becoming is interactive between the mother and the offspring-to-be; it is a powerful transaction that merits a private sanctum not to be rudely or casually invaded.

Is the single zygote cell that begins the profound transformation human? Scientifically there is no question; the zygote is certainly human to its core. But does the zygote display all of the characteristics of a human being? Can a single cell be a human being, a person, an entity endowed with unalienable right to life, liberty, and the pursuit of happiness? This question cannot be answered on scientific grounds alone. Human being, person, and human rights are not terms stemming from scientific definition. But one trained and accustomed to think as a scientist cannot fail to note significant disparity between the common meaning of these terms and a single cell.

When we view the public world of real people dealing with

each other in the complex and subtle interactions of social life, it seems ludicrous to suggest that concepts appropriate to that realm should be extended to an individual cell at the bare limit of ordinary visibility. It makes as much sense as declaring acorns to be oak trees and selling them at oak tree prices. Less ludicrous, but much more difficult to answer, are questions about what should, in fact, be the status of the zygote and—most difficult of all to answer—exactly when in the course of development full human status should be assigned.

Controversy over these questions continues to agitate many minds—of presidents, justices, and legislators as well as physicians, scientists, lawyers, philosophers, and theologians. No answer has yet proven universally persuasive; the questions clearly pose a number of imponderables. Having been unresolvable to the full satisfaction of all so far, the questions toss in a sea of uncertainty and debate, leaving the unborn without clearly defined status. As one consequence, medical effort to improve treatment of the unborn is hampered and growth of scientific knowledge about them is constrained.

In democratic society, status confers rights protective of the interests of individual members. In the United States, equality before the law is constitutionally guaranteed and every adult citizen has the right to vote. Internationally, freedom from deliberately inflicted pain and suffering is the subject of solemn covenants not always observed. Such human rights are aspects of status, especially in countries that vest fundamental value in each individual human being. Status may change in the course of life, as in acquiring the right to drive at the age of assumed responsibility or losing the right to be free on conviction of a felony.

Therefore, if the status of the unborn is unclear in national jurisprudence, their rights are also ill defined. In the United States, the situation is part cause and part effect of dissention as to how to treat the unborn in the politically tense area of abortion. The tension has infected consideration of other reproductive issues, including such innovative technologies as external or *in vitro* fertilization (IVF) and fetal research in general. The tension and dissension also have widening effects in the whole area of public reproductive health policy.

While there is no simple answer as to why this tension has come to pass, in no small measure it is because the unborn have for so long been cloistered within the maternal womb where they are not accessible for direct interaction with others. Even when the mother becomes big with child, the fetus itself remains unperceived.

The normally covert existence of the unborn is, of course, derived from our origins as mammals. All mammals reproduce by internal fertilization and temporarily nurture their offspring within the uterus of the mother (gestation or pregnancy). This arrangement can be interpreted as an advantage in human evolution because it entails minimal conscious attention, proceeding largely through involuntary mechanisms. Except in very late stages or in abnormal cases, the pregnant woman can go about other duties with little or no handicap.

These mammalian facts of life have profoundly shaped values and customs in all human societies. They have particularly shaped the role and lives of women, all of whom carry the mark and the burden of this mammalian heritage. It therefore comes as no surprise that women's rights to reproductive choice, especially with respect to decisions about the maintenance of pregnancy versus abortion, mobilize the efforts and concerns of activist feminists.

Moreover, it follows that during the close mutual interaction of mother and developing offspring in pregnancy the status of the unborn is difficult to separate from that of the mother. For some time neither the status nor welfare of the one can be altered without affecting the other. This has tended even further to suppress attention to the status of the invisible "insider" who

7

is at first little more than a part of the socially much more firmly established "outsider."

In recent years, however, the situation has been changing noticeably in several significant respects. One example is a fact that increasing knowledge and altered reproductive practices have taught us. The earliest phase of human development, while it normally occurs in the oviduct of the mother, is actually quite independent of her. In the IVF procedure these same stages proceed quite normally outside the mother and in an artificially prepared solution that is incubated in a suitable gaseous environment. Clearly, early dependence on the mother is actually relatively minimal in that it is readily satisfied in other ways.

This phase of relative independence begins at ovulation, when the egg is discharged from the nurturing ovarian follicle and starts its journey down the oviduct or fallopian tube into the uterus (see figure 3).* If the egg meets sperm in the oviduct and is successfully fertilized, in a few days the developing entity— the preembryo—reaches the uterus and implants in its wall. Implantation may be thought of as the physiological beginning of actual maternal pregnancy as well as of the offspring's significant dependency on its mother. Until that time the developing entity is within the "mother" but not importantly interactive with her, as the following facts indicate.

First, the woman within whom the egg is fertilized is normally unaware of when fertilization occurs. The union of the male and female gametes is a complex and fascinating story in its own right, but there are no signals to either parent as it happens. Moreover, there is no discernible effect on maternal functions until about a week later when the first detectable hormonal

---

* Figures have been grouped in an appendix at the end of the book. Readers who are unfamiliar with the normal process of human development may find it useful to go through the appendix at this point to familiarize themselves with its general nature and the terminology used for descriptive purposes.

changes occur, in association with implantation of the preembryo in the uterine wall. If implantation does not occur, which is perhaps half the time in the human species, the woman never shows any sign of pregnancy, nor does she have any knowledge or recall of the temporary presence within her of a potential offspring. It follows that there is no easy way to assign status of any kind to so transient an entity when its very existence is unrecognized.

In terms of maternal effect, therefore, pregnancy begins not with conception but with implantation. Measures that inhibit implantation (the intrauterine device or the "morning-after" pill) are more accurately called contragestive (antipregnancy) than contraceptive (anticonception). This does not change the fact that both block production of an offspring resulting from sexual intercourse, the objective of all methods of birth control.

Second, as noted, the conditions necessary for preembryo development external to the mother are readily supplied in the laboratory, as is done in the IVF procedure. They involve only incubation at controlled temperature in a relatively simple physiological solution. However, this creates an entirely new situation with respect to status. The preembryo is now fully revealed, there is no uncertainty about its existence. On the contrary, we shall see that very real questions and issues of status arise in connection with it—quite separately from those that arise in connection with abortion.

Third, also as already noted, in the normal process within the mother it is not until implantation that hormonal changes announce the beginning of the pregnant state. Roughly concurrently, complex interactions and exchanges between the developing entity and the maternal circulation are initiated. With these events independence of mother and offspring is terminated and a relationship between two living entities begins that is as complex, intimate, and complete as any known to biological science, or to ethics and philosophy.

Fourth, the independence of the preimplantation phase is dramatically demonstrated in the few IVF cases in which an externally developing offspring has been transferred to the uterus of a woman who is not its genetic mother. Development continues in such a substitute gestational mother. Thus *genetic* and *gestational* motherhood can be separated, and the potential developmental history of the early offspring can be realized totally independently of the genetic mother.

Fifth, preservation by freezing (cryopreservation) further extends the gestational independence of external preembryos. At least theoretically they might be maintained indefinitely in the frozen state, even well beyond the life of the genetic mother. Who, then, is responsible for their welfare and their fate?

The five circumstances outlined indicate that the earliest phase of human development can occur separately from and independently of the genetic mother. Under some circumstances, therefore, the status of this phase must be addressed as an entity physically separable from its mother. The point is dramatized and becomes pressing with the practice of frozen storage, through which the preembryo may survive even beyond the death of both parents.

In the early stages of development of human IVF (in the late 1970s), pregnancies and births were obtained but at unsatisfactorily low rates. However, when the earlier developed technique of superovulation was applied, success rates were significantly raised. Superovulation involves pharmacological stimulation of the woman's ovaries so that they produce more than one egg in a given month. Several eggs can then be recovered, fertilized, and transferred to the uterus, increasing the chances that at least one will implant.

But such stimulation of a woman's ovaries often produces an embarrassment of riches—four or even more eggs. Transfer of all of them not only increases the pregnancy rate but also the frequency of multiple pregnancy. In turn, multiple pregnancy, particularly more than two, elevates obstetrical risk to both mother and offspring and imposes rearing burdens that understandably daunt many couples.

The freezing of "surplus" preembryos, beyond the several that seem optimal for transfer, therefore offers possible advantages. For example, it allows storage of the surplus to await a possible second transfer attempt if the first fails. However, in the early trials it quickly became apparent that frozen storage also heightened and sharpened issues raised by the uncertain status of preimplantation stages. In one instance, at an Australian infertility center in Melbourne where frozen storage was first practiced successfully, two frozen preembryos were stored pending success or failure of a first attempt at transfer to a patient's uterus. Unfortunately, before a second attempt was made, the American donor couple was killed in an airplane accident, thus inadvertently abandoning and "orphaning" their two "offspring." How to dispose of these offspring became the subject of intense debate in Melbourne and in the international press. The incident significantly affected the course of IVF legislation then pending in the Australian state of Victoria where Melbourne is located, and the case remained unresolved some five years later.

The crux of the issue, of course, is that cryopreservation results in reversible suspension of life activities, essentially a pause in biological time. So far as is known, the state of suspended animation can be maintained indefinitely without significant change in the frozen material. Whatever the status, or its lack, prior to freezing, it too remains frozen. The result is a classic dilemma. The longer the passage of time, the more attenuated becomes the connection between the static offspring and the aging genetic parents.

Indeed, the same kind of disparity between siblings has been reported by the press from England. A first transfer was successful but frozen preembryos from the same "litter" were used in a later, also successful, transfer. Although the two offspring were of the same age in time elapsed from fertilization (in this sense fraternal twins), they were several years apart in developmental age (and thus like ordinary siblings).

Such situations offer a number of speculative issues that will be returned to later. In the Melbourne case, the status question became acute because the genetic connection with the parents as gamete donors was abruptly interrupted. The question then arose as to who has the right or responsibility of decision regarding what happens next.

The dilemma is not made easier by the opinion, strongly held by right-to-life community leaders in Melbourne, that to discard potential offspring, whether prior to freezing or after, would be a wrongful violation of their fundamental right to continued life. This view stems, of course, from the contention, forcefully advanced in the abortion debate, that an individual life begins at conception (fertilization). Almost as if it were contrived to do so, this contention underlines the ambiguous and unresolved status of the preimplantation stage of human development.

Thus the pro-life contention sidesteps the difficult and delicate questions of appropriate status at particular stages, as well as in particular circumstances, by declaring that a person exists fully and absolutely from the first initiation at conception. Such a definition precludes induced abortion at any time or under any circumstance, which is the objective of the right-to-life movement. However, it also places severe constraints on other reproductive options such as IVF, which seeks to generate life rather than to terminate it.

This inevitable policy linkage of a right to life that begins at conception with limitation of reproductive options was made crystal clear in the recent document issued by the Vatican's Congregation for the Doctrine of the Faith.[1] It is an instruction to members of the Catholic faith that reasserts a rigid doctrine in the face of advancing reproductive knowledge and its derivative technology, even though the technology is designed to yield a clear human benefit.

Moreover, the central issue, as the Vatican document perceives it, is not only the moral but the *public* status of the unborn at all stages, including the preembryo. The document requires brief quotation to convey both its substance and its flavor. It deals successively with "fundamental principles," "respect for the human being," "moral questions raised by technical interventions on human procreation," and "the relationships between moral law and civil law."

Under fundamental principles, the Vatican document notes that "thanks to the progress of the biological and medical sciences, man has at his disposal ever more effective therapeutic resources; but he can also acquire new powers, with unforeseeable consequences, over human life at its very beginning and in its first stages. Various procedures now make it possible to intervene not only in order to assist but also to dominate the processes of procreation. These techniques can enable man to 'take in hand his own destiny,' but they also expose him 'to the temptation to go beyond the limits of a reasonable dominion over nature.' They might constitute progress in the service of man, but they also involve serious risks."[2]

To cope with the risks, "science and technology require, for their own intrinsic meaning, an unconditional respect for the fundamental criteria of the moral law; that is to say, they must be at the service of the human person, of his inalienable rights and his true and integral good according to the design and will of God. . . . science without conscience can only lead to man's ruin."[3]

The caveat applies especially "in the field of sexuality and procreation, in which man and woman actualize the fundamental values of love and life. . . . Such values and meanings are of the personal order and determine from the moral point of view the meaning and limits of artificial interventions on procreation and on the origin of human life. These interventions are not to be rejected on the grounds that they are artificial. As such, they bear witness to the possibilities of the art of medicine. But they must be given a moral evaluation in reference to the dignity of the human person, who is called to realize his vocation from God to the gift of love and the gift of life."[4]

Human procreation differs from all other. "The transmission of human life is entrusted by nature to a personal and conscious act and as such is subject to the all-holy laws of God; immutable and inviolable laws which must be recognized and observed. For this reason one cannot use means and follow methods which could be licit in the transmission of life of plants and animals."[5]

With respect to "interventions upon embryos and human foetuses," the Vatican document draws a major conclusion: "The human being must be respected—as a person—from the very first instant of his existence. . . . Life once conceived, must be protected with the utmost care; abortion and infanticide are abominable crimes. . . . Human life must be absolutely respected and protected from the moment of conception. . . . This teaching remains valid and is further confirmed, if confirmation were needed, by recent findings of human biological science which recognize that in the zygote resulting from fertilization the biological identity of a new human individual is already constituted. . . . Thus the fruit of human generation, from the first moment of its existence . . . demands the unconditional respect that is morally due to the human being in his bodily and spiritual totality."[6]

The position taken by the Vatican document, with passionate finality, sharply poses certain central questions in setting public status for the unborn, particularly in communities containing significant Catholic or other groups whose reproductive morality proceeds from religious doctrine or other fixed tradition. What are the purposes to be served in a pluralistic community by *public* policy as against group or personal morality? Should any *particular* morality be made the basis of public policy? What are the appropriate targets for public policy? Can the impacts of a formulated policy be confined to consensually selected targets or will policy have unintended consequences beyond its target area?

Given the complexity and profound importance of reproduction in human society, it seems clear that assignment of status to any particular stage of reproductive realization is likely to have multiple and complex consequences, some of which may be foreseeable and intentional; but others may not. To assign status rigidly on the basis of a morality subscribed to by only a fraction of the population will seem arbitrary to all others and will only invite continued dissention. The purpose of public policy should be acknowledged to be limited and limiting—that is, to set consensual limits within which particular moralities can be practiced with respect and tolerance.

In the matter of abortion, as a case in point, many people regard the procedure as regrettable at best but nonetheless justifiable under special circumstances—for example, pregnancy resulting from rape. Such persons, who might favor contraceptives and contragestives as a means of family planning, might strongly reject abortion for the same purposes, particularly if the intention were to select the sex of offspring by aborting the undesired sex. Rigid right-to-life definitions of unborn status, as exemplified by the Vatican document, exclude realization of all of these purposes—a result certainly not generally acceptable in the pluralistic ethos of the United States.

Thus a general issue is joined by the Vatican document: Can status of the unborn be effectively defined for contemporary purposes if the status is founded in absolute and rigid criteria that are largely derived from past tradition and incorporate assertions of faith that appear to take little account of the findings of contemporary developmental science?

For example, contemporary knowledge has revealed that there is no *instant* of conception; the *process* of fertilization (or syngamy) extends over many hours. This precise fact has become an issue in another continuing controversy in Melbourne with respect to the legitimacy of surgically introducing a sperm into a human egg as a possible alleviation of male infertility. The issue is whether the actual intervention occurs *prior to* conception and is, therefore, allowable. One cannot help recalling the medieval issue of the number of angels able to dance on the head of a pin.

We shall see in later chapters that, scientifically, conception is only a partial boundary (albeit a significant one) when one views the entire continuous and complex process of genesis of a person. Beyond even that question, however, definitive founding of so central a matter as social status on a single biological aspect fails to take into account the varied and often subtly different social contexts within which the concept of person is applied. Rigid and overly simplistic criteria may seem gratifyingly final, but their simplicity often proves illusory when confronted by complex reality.

What we require are status criteria based in *contemporary* knowledge of the nature of developmental change combined with sensitive appraisal of specific social circumstances and contexts, such as are invariably associated with decisions as grave as abortion or assignment of rearing responsibility for a child. The general policy dichotomy thus presented is between *absolute* criteria based on traditional concepts and *relative* criteria appropriate to the current state of knowledge and *specific* currently formulated purposes.

Status based on relative criteria obviously offers greater flexibility and wider opportunity for accommodation of differing views—both of which are advantages for reaching consensus in a pluralistic society. To the degree that consensual dominant objectives can be articulated by such a society, assuming the society to be committed to mutual tolerance, status can then be crafted politically to meet even a diverse set of consensual objectives.

It is worth recalling in this connection that the current status of adults has had a long and continuing history of contention—social, political, and even military (consider slavery, doctrines of racial and ethnic superiority, the relative status of the sexes, and the violence of religious wars). Fortunately, processes of negotiation and accommodation have been able to resolve at least some of these status issues. The same processes need now to be applied assiduously to the status of the unborn.

What realistic opportunities are there to achieve some degree of accommodation with respect to the status of the unborn? How might an accommodative process be set in motion? What current consensus exists that might provide a launching platform for such an effort? Several objectives that have been articulated and have reechoed in the abortion debate and in discussion of new reproductive options deserve consideration as a possible accommodative base.

First, and probably foremost, is the widely expressed need to preserve the special quality and dignity of human life. Shared in general by most persons, it is the most strongly pressed objective of the right-to-life position. As it is often formulated in religious terms, it cannot be incorporated directly into U.S. public policy without breaching the fundamental and strongly supported doctrine of separation of church and state. However, it is in fact also a primary secular value in the United States, an objective that permeates our national philosophy—as do individual rights and freedom. Suitably framed, a consensus position assuring respect and dignity to the human quality of the unborn might easily be included in their status assignment.

Second, many people are uneasy about the increasingly powerful and precise ability of science and medicine to intervene in and modify natural human reproduction. The uneasiness is over whether this capability may advance more rapidly than our wisdom in employing it—that it may be used either inadvertently or deliberately to diminish human beings. This concern could be made an accommodative objective in considering such questions as: How should the principle of informed consent, as developed by adults for adults, be translated to fit the unborn who are not yet capable of deliberation or decision about their own fate? (The question is not a new one, of course, but it is increasingly applicable not only to the now arising generation but to the subsequent generations that will be their offspring.) Given the possibility of even a limited impact on future generations, what interventive reproductive options are sanctionable and which ones should be constrained? How can the range of tolerable intervention be delimited without excessively confining advancing knowledge and practice?

In general, I believe that in assigning status to the unborn, consensus should be sought to recognize and protect their critical human potential, both immediately and for the longer-term future. The unborn should be valued not only as offspring but as ancestors of generations-to-be. Policy decisions about them must therefore enlist the highest standards of human concern and wisdom.

Third, there is general concern about whether and when the unborn may suffer pain and discomfort. It should be a shared objective among contending points of view to better understand the developmental maturation of pain mechanisms and to apply that understanding to definitions of status that leave margins of safety when there is uncertainty. More will be said of this in later chapters.

Fourth, general support exists in the United States for the advancement of knowledge and the expansion of technology as means to further health and control disease, providing that the process does not threaten other major objectives. The motivation is perhaps strongest among health professionals. Some nonprofessionals are concerned that the strength of personal professional motivation may sometimes outweigh other important considerations, particularly in connection with research on human subjects. This is an important factor in the debate over status of the unborn. To deal with these matters, a national consensus statement might be developed to assure commitment to increase biomedical knowledge and its application, balanced against specified essential considerations such as informed consent and avoidance of imposed suffering on human and other sentient beings. Such a statement might reduce existing tension and distrust and facilitate accommodation and realistic consensus on status of the unborn.

Finally, the prospect of what has been called human engineering generates concern about scenarios similar to the Brave New World. With respect to the status of the unborn, the thrust of such scenarios is illustrated by the question of whether, given the current capability to produce human embryos externally, the technique might be used in future "people farms" to create human embryos solely for experimentation, as a source of cells and tissues for transplantation therapy, or to provide an underclass of slaves and servants for a dominant stratum of aristocrats. It would be a useful objective to develop consensual guidelines on such prospects before they become possible. The general presumption might be that interventions practiced on human beings at any stage are unwarranted unless *specifically sanctioned* for consensually derived purposes through a formal process of decision which is accessible to all interests.

This list of possible consensual objectives is substantial but is not intended to be exhaustive. Rather it is a sample, to be modified and expanded during a broad and continuing deliberation. One generalization from the sample is that specified concerns and objectives should be formulated first; then they can be the targets for specific proposals for status definition. The proposals should be evaluated in relation to their effects on their targets, not only in terms of the expected mitigating effects but for their possible unintended and unanticipated effects as well. Moreover, the political feasibility of the proposals must be examined along with their probable effects. Nothing will be gained by defining a status that cannot be implemented because of unexpressed but deeply entrenched moral objections.

Also, the defined status must be realistic and not overly ambitious in objective. In the abortion debate, right-to-life and right-to-choice groups have rallied to their opposing slogans because they believe that most people see both continued life and free choice as desirable objectives. But neither indefinitely continued life nor unlimited choice are assured even to adult persons. To try to assure both in the intimate conjunction of pregnancy necessarily sets the stage for conflict.

More modest and less confrontational concepts and language are needed for accommodative solutions, especially since the phenomenon of pregnancy involves what may be the ultimate in human mutuality—for both the mother and the unborn fetus. The concept of status for the unborn must not only placate the concerns of individuality but assure the benefits and value of the profound mutual experience of pregnancy—experience that locks into place one of the most powerful human bondings, that between mother and child.

In the next chapter we shall look more carefully at the matter of human individuality itself. Personal individuality is very closely linked to the sense of self—and the security of self is a persistent and fundamental individual concern. Given this crucial concern, the origin and genesis of the individual self must be addressed fully in formulating the status of the unborn.

CHAPTER 2

Becoming
an Individual

IN western liberal tradition, society is made up of individual persons who are its members, contributors, and beneficiaries. Individuals enter into complex relationships within society that, ideally, serve the interests of both the individuals involved and the society as a whole. These relationships are governed, in significant part, by what we are here calling status, whether established by law, tradition, or shared customs and habits.

Therefore, in considering what status to assign to the unborn —and when—a foundational factor is the nature and degree of their individuality. This is not a question normally raised about adults; adults are the exemplars of individuality, and only when their normal activities and capacities are gravely impaired does the question arise. But the unborn do not begin as individuals in the usually accepted social sense. Rather they begin as a single cell, a phenomenon and conception that is entirely biological and not at all social in its meaning and connotation.

In the complex processes of biological development, when and how does human *individuality* arise? What relationship is there, or should there be, between what happens in terms of biological individuality and what comes later to be accepted in terms of social individuality as status? These are difficult questions, and I shall not attempt to confront them here all at once. Rather I shall attempt to approach them gradually, much as individuality in the full human sense emerges in the course of development.

Meanings of "Individual"

In the most general terms, to be an individual is to be a single and particular entity, therefore recognizable as *both* unitary and unique. Unitary refers to singleness and unique refers to being particular in the sense of having an identity distinguishable from others. The total character of one individual is recognizably different from that of another, even though they are similar (and even if they are identical twins, which is a special case).

It is important to recognize the two distinct meanings of "individual"—to be one entity and to be unique—because the two properties do not arise simultaneously in the process of becoming. Neither do several other aspects of individuality that we take for granted in discussing the status of adults.

In this chapter I point to six separable aspects of individuality that arise during a life history: genetic, developmental, functional, behavioral, psychic, and social. Genetic individuality refers to hereditary uniqueness, where "hereditary" means "able to be transmitted from generation to generation"; developmental individuality refers to achievement of singleness and its consequences; functional individuality refers to diverse activities essential to survival; behavioral individuality refers to integrated activities of the whole in relation to environments; psychic individuality refers to inner experiences accompanying behaviors;

and social individuality refers to self-aware interactions within a community of individuals.

Each of these aspects of individuality arises in the process of becoming an adult human being, but they appear at different times. The question thus posed is which of these, or what combination of them, merits new status?

Before addressing this central question, I should reformulate it as it is frequently asked in connection with abortion: When does life begin? The motivation of the question in this form is the assumption that once human life does begin, it may not be terminated until it runs its natural course; it is protected so as to be terminable only by higher authority—whether societal or divine.

To be answered scientifically, the question needs further definition before it can be addressed. Human life, in one sense, has no beginning in our time; it has existed continuously on earth since its inception millennia ago—whenever and however that is assumed to have occurred. In this, human life is like all other life—it has been transferred in unbroken successions of generations since its origin. No life begins on earth today except from ancestors; life invariably passes from one generation to another similar to it.

Without disputing the fact of continuing life, some will protest that this is not the point of the original question. Certainly, they will say, something new does arise in each generation; each individual does not extend back in time to the beginning of humanity. This illustrates the second sense in which the question and the term life can be understood. The question then is: When does a particular individual life begin? That is a different and more difficult question, stressing individuality. It is the crux of the issue to be discussed in this chapter.

As earlier mentioned, right-to-life adherents seek to simplify the problem by declaring that an individual begins at conception; some urge that such a simplification should be written into the U.S. Constitution. Scientifically this would be egregious oversimplification. To see why, one must look more deeply into what fertilization or conception (syngamy) actually signifies with respect to individuality.

Conception is not a technical term, and it has a variety of meanings in ordinary usage, as in conception as a mental construct. One of its other meanings is biological and synonymous with fertilization, the union of male and female gametes (or syngamy). Fertilization has two major consequences: It activates what has been a dormant state of the egg in the ovary, so that it completes its maturation and continues development; and it combines hereditary contributions from both parents into a new and unique hereditary constitution (genotype or genome).

The genome, as we now know, is written in the chemical language of macromolecules, specifically the nucleotide sequences of DNA. The genome governs the hereditary aspect of the subsequent developmental course. It does this because exact genomic copies are transmitted to each new cell generation and because all new materials made in every cell reflect the molecular genomic message. The process involves expression of the genotype, ultimately to produce the phenotype—the characteristics of the fully manifest new individual.

Thus fertilization substantially alters the properties of the egg, not only by activating it but also by providing it with a new joint genome derived from both genetic parents. What does this mean in terms of individuality? While activation is essential to everything to follow, it makes little immediate contribution to individuality. It takes an expert observer to detect that fertilization has occurred during the first few hours after sperm penetration. But something profound has happened nonetheless, although it will require considerable time before it is visibly displayed.

The fertilized egg (zygote) has acquired a very important first aspect of individuality in its unique genome, which combines contributions from both parents but does not duplicate either of them. The new genome is different not only from that of either parent but also from that of any sibling (other than an identical twin). On the other hand, the new genome shares by far most of its characteristics with those of its parents and siblings—the basis for its human hereditary lineage and kinship. This genetic aspect of eventual total individuality is, in fact, a linear product of all previous generations and will influence all that is hereditary in this generation and beyond. The new genome is, therefore, a fundamental first step in establishing a new total individuality that will progressively emerge.

But, even while recognizing the profound importance of this first step, it is essential to keep in mind how much is yet missing at this early stage from what will be present when full human individuality is achieved. Uniqueness in the genetic sense has been realized but, for example, unity or singleness has not. The zygote, with its unique genome, may give rise in either natural or induced twinning to two or more individuals with identical heredity. In some species this occurs naturally and regularly. Moreover, in mice (and probably in most if not all mammals including humans) cells of two or more different genotypes can be combined to form one embryo which develops into an adult that is a mosaic of more than one genotype. In humans, on the other hand, identical twins with the same genome have their own sense of individual identity, recognized by different names and separate voting rights.

To misidentify initial genetic individuality with full individuality harks back to preformationism, a theory that was entertained more than a century ago. In those scientifically less well informed days, observers, using primitive microscopes, thought they saw a tiny miniature of a person packed into the egg or even into the much smaller head of a sperm. All that was necessary, they thought, was to enlarge these miniatures through development to yield people. If that were true, one would not have to figure out how complexity comes into existence in each generation. It would be there all the time in miniature.

Today we know that miniaturization is too simple a concept for hereditary transmission. What is transmitted between generations is a message *coded* in DNA. The message has to be replicated in the form of DNA, transcribed as RNA, translated into amino acid sequence in proteins, and then elaborated in a multitude of ways in order to reestablish the complexity of a new individual in the full sense. The new genetic message established at fertilization is unique, but it is still far from realization in a particular human adult individual.

We know further that genetic individuality is a separate property from developmental individuality (recognizable in singleness) because though genetic particularity or uniqueness is established at fertilization, singleness arises independently and at a significantly later time.

The Origin of Singleness

When does human *developmental* singleness first appear? About ten to fourteen days after fertilization and by a process still only dimly understood more than a half century after it was discovered as a fact. At the time this second aspect of individuality appears, cell division has produced perhaps a hundred cells and fluid has accumulated within the cellular mass to produce an eccentric inner cavity (see figures 1, 2, and 3). It is this transitional stage, referred to as the blastocyst, that arrives in the uterus and begins the process of implantation into the uterine lining.

At this stage, the neighboring cells of the blastocyst are closely adherent rather than loosely associated as in immediate postfertilization stages. The blastocyst, therefore, is no longer merely a bundle of cells, it is a multicellular entity. Moreover, within it there are now two distinguishable cell populations, those of its external layer and those of a small cellular mass extending into the inner cavity. Only this inner cell mass is the precursor of the embryo; the external cells and their descendants become the placenta and extraembryonic membranes, which are discarded at birth.

So far as is known, the cells of the inner cell mass of the early blastocyst are little different in developmental capability from the zygote. Each can contribute to any part of the embryo, and separation of the mass into two parts can still yield two or more embryos. It is only when the later-stage blastocyst has penetrated and implanted in the uterine wall that properties of the inner cell mass change and it becomes committed to the production of a single individual.

The stage of commitment to developmental individuality is often referred to as primary embryonic organization—the beginning of formation of the embryo proper. In the process, the inner cell mass is transformed into an embryo—a rudiment capable of building a single and complete multicellular individual. In this sense, primary organization lays down the basic structure of a single organism. It is represented in the three main body axes: head and tail, left and right, back and front.

The first sign that primary organization is underway is the appearance of what is called the primitive streak (see figure 4). This is a linear thickening, visible with minimal magnification, that lies in the head-to-tail axis of the embryo-to-be. The embryonic structures of the head-to-tail axis first appear ahead of the primitive streak in terms of later relationships (see figure 4A, B, C). Subsequently, the axial structures seem to spin out, from head to tail, as the streak is displaced backward (see figure 4B).

It is to be emphasized again that before these primary organizing events occur, there is no recognizable embryo, only a two-layered disc of cells that is precursor to it. *From the time of appearance of the streak,* singleness of future development as an organism is being established. If two streaks appear or are induced to appear, two embryos form. The complex and fundamental changes involved in primary organization, while still not fully understood, clearly entail the onset of developmental individuality in the sense of singleness.

Organogenesis

The events that follow first block out the major regions and organs of the body and then fill them in with ever-greater detail. In so doing, what begins as the unity of developmental individuality is differentiated into specialized diversity and this, in turn, is later once again integrated into higher-level *functional* individuality. The process of diversification into parts as organs is known as organogenesis, the major activity of the embryonic period. Organogenesis continues until about the end of the eighth week, although this conventional boundary is more a matter of convenience than of sharp discontinuity. Organogenesis, though declining, actually continues well beyond eight weeks in particular organs and parts, for example, in the formation of teeth.

As organogenesis proceeds new structures and properties come into existence, and these provide new activities and functions. Moreover, as more mature functional activities are realized, the offspring's potential for existence independent of the mother is increasing. This rising potential is manifested as increasing *functional individuality.*

A good example is provided by the development of the heart. Not surprisingly, given the importance of effective transport of materials to and from the growing and developing embryonic organs and tissues, the heart is the earliest organ to become functionally active. Its beat begins during the fourth week after fertilization.

In its earliest functional state, the heart is a simple tube, with only two instead of four chambers and no valves. In this primitive form it can only move blood back and forth rather than in a continuous circuit through the developing network of embryonic circulatory blood vessels. However, soon the heart, blood vessels, and placenta mature sufficiently to yield a true one-way circulation, thus assuring effective nutritional and respiratory exchange between the mother and her offspring (see figure 9). At the same time, the embryonic heart and circulation are establishing a foundation for independent external life after birth.

As it is with the heart, so it is with other organ rudiments (see figures 5 and 6). Each appears, grows, and acquires greater definition, while differentiating, maturing, and achieving increasing effectiveness of function. Significantly, as noted, the heart and circulation appear early. This allows them to meet the requirements of all other organs and parts for nutritional and other exchange with the maternal organs. Similarly, although the nervous system as a whole reaches function later than circulation, its formation likewise begins very early in development. Like circulation, neural communication serves the need for functional interaction among the specialized organs of the increasingly complex organism. We shall see in the next chapter that neural communication through the spinal cord and brain stem, which is necessary to basic functions for survival, begins much earlier than the more complex communications involving higher levels of the brain.

Closely associated with the development of the nervous system is the appearance and maturation of the major skeletal axis and its associated musculature (see figures 4C and 5). These are not only structural elements for emerging body form but the major foundation for later *behavioral individuality*. This is especially apparent in the limbs, which first appear as formless buds at about five weeks, are paddlelike at six weeks, and have extended into recognizable arms and legs by eight weeks (see figure 5B, C, D). Musculoskeletal maturation is closely linked to that of the nervous system. Together, by around the eighth week, they have given rise to sufficient integrated functional individuality to be displayed in primitive behavior. The end of the eighth week has traditionally been regarded as the transition from the stage of the embryo to that of the fetus.

An important point should be emphasized here. Specific structures must be laid down before specific activities can be manifested. Therefore, as organogenesis advances structurally, it sets the stage for increasing functional individuality, including behavior.

However, level of function cannot always be inferred from structural characteristics alone. Although by eight weeks after fertilization structural maturation has advanced far enough to make some external features of the embryo recognizably human, structural maturation internally is still quite incomplete and many functions are correspondingly limited. This important caveat against judging inner states from external appearance alone is especially applicable to the central nervous system, as will be made clear in the next chapter.

Behavior as Function

Behavior can be considered to be a special aspect of function. If function is activity essential to maintain the integrity of the organism, behavior is function that relates the organism to its environment. Movement is an especially prominent and characteristic aspect of all animal behavior, including that of humans. In its most obvious form it depends on functional neuromuscular maturation. This means not only that muscles and nervous system must each be sufficiently mature but that the interactions between them must be as well.

Some fifty years ago, human embryonic movements were

studied as the earliest manifestations of *behavioral individuality*. The observations were made on embryos delivered in the course of therapeutic abortion. The embryos were placed in a warm and physiologically balanced solution, to replace the amniotic fluid in which they are normally immersed, for brief observation prior to their inevitable death.

Such early fetuses were found to respond weakly to gentle stimulation around the mouth by turning their head away from the stimulus. At the developmental age of seven to eight weeks, a fetus would slowly turn its head, a movement involving contraction of appropriate neck muscles. The movement was interpreted as an avoiding (aversive) response to stimulation. With increasing levels of maturation, fetuses gave stronger and more varied responses. All of the responses were of a kind that behavioral scientists call reflexive—that is, the responses were stereotypic and repetitive, with no evidence of modifiability with continued stimulation.

Detailed anatomical studies revealed that by seven to eight weeks the neck muscles involved in turning of the head were sufficiently mature to undergo contraction. Moreover, these muscles were found already to be innervated—nerve fibers terminated on the surfaces of the differentiated muscle cells (see figure 12). The nerve fibers were extensions of neurons whose cell bodies were in the central nervous system, in the so-called ventral horn. This is the area in adults where motor neurons lie, neurons that, if stimulated, cause peripheral muscles to contract.

In turn, these neurons innervating the neck muscles were in neurological contact with other neurons that had processes extending peripherally to sensory endings around the mouth. Thus, at the stages when the early behavior of neck turning appeared, there was a structural and functional neurological substrate in place—a series of neurons connected the apparent sensor in the area of the mouth to a motor responder in the neck. In more general terms, there was a demonstrable neurological pathway or circuit to account for the behavioral phenomenon (see figure 11).

In recent years, computer-assisted imaging with ultrasound has been used to confirm and extend the earlier direct behavioral observations. The technique depends on reflectance of high-frequency sound waves, together with sophisticated electronics and a television-type monitor. It can penetrate into the abdomen, without so-far detectable harm, to yield real-time images of the embryo within the uterus. A trained observer can interpret levels and normalcy of organ development as well as record movements of the embryo or fetus over extended periods. Because of the clinical usefulness of the data obtained, a number of studies of fetal movements have now been recorded.

Significance of Movement

It is now certain that the reports of early movements of aborted fetuses, even though the fetuses were in a moribund state, substantially correspond to behavior of normal fetuses in the uterus as well. Ultrasound imaging confirms fetal movements of essentially the same kind and at essentially the same stages, beginning irregularly and infrequently even as early as the sixth week. A more complex startle response has been reported in the seventh week, and both general movement and such special movements as hiccup and various isolated limb movements occur in the eighth week. By the ninth week hand movements to the face and sporadic breathing movements have been reported. In the tenth week yawning occurs, and in the eleventh week swallowing and sucking have also been observed.

A major difference from the earlier studies, however, is that all of these movements appear to occur in the *absence* of known stimulation. "Spontaneous" movement raises the question of whether there is inner volition and even possible sentience. These are assumptions treated as certainties in the much-publicized descriptions and interpretations highlighted in the videotape *The Silent Scream*, based on ultrasound recording of early fetal movements and widely circulated to generate anti-abortion sentiment. More will be said about its content and interpretation later, including the problems involved in drawing such conclusions (see chapter 6). For the moment, movement is cited only as representing the onset of behavioral individuality. In its significance as a stimulus to human interactions this is a new dimension, important not only in relation to social status but needing careful consideration in relation to the possible beginning of *psychic individuality* as well.

Emergence of Psyche

What is meant by psychic individuality? I use the term here to designate inner subjective experience, such as each adult is directly aware of, testifies to as a common accompaniment of behavior, and, based on the behavior of others, assumes to exist in them as well. From this definition we can proceed to the common assumption that a fundamental characteristic of human beings is their inner sense of being that is variously designated as sentience, self-awareness, consciousness, or, more generally, psychic individuality. This term underlines the uniqueness and particularity of inner awareness as it is experienced by individuals designated as persons. The critical question then is when, in the course of development, a person first exists to experience such individuality in at least minimal form.

The major problem in considering the question is how to identify inner experience in a being that cannot communicate through some form of behavior, whether the being is a prematurely born infant in an intensive care nursery or a fetus in the uterus. Still more troublesome is the possibility that, in a primitive state, such individuality may exist as self-awareness in the form of only unresponsive listening or other sensing without accompanying motivation or capability to communicate.

Such questions are obviously difficult to answer. At the moment, in the absence of any identifiable physical sign such as specific electrical activity of a particular brain site, we have no choice but to rely provisionally on some kind of communication as a primary means to recognize psychic individuality. However, the difficult questions must be kept in mind, because they emphasize that the absence of behavioral signals from a fetus cannot prove that it has no inner experience or psychic individuality.

The uncertainty is further emphasized by recently reported studies of two adult patients suffering from a disorder known as prosopagnosia, a condition associated with bilateral damage to the occipito-temporal region of the brain cortex.[1] During testing, the patients do not report that they recognize faces well known to them from previous experience—for example, faces of close family members or famous people. However, when they are shown these pictures, there is a detectable electrical change in their skin, just as there is in normal people.

This fact has been interpreted to mean that the patients do, in fact, respond to facial familiarity even though verbally they do not report awareness of it. Even in adults, therefore, response and awareness are not necessarily fully correlated. In the fetus, too, correlation of the two might be absent—with awareness but no evident response.

However, it is also true that psychic individuality, in the sense of inner awareness such as ours (reports of disembodied spirits notwithstanding), is not known to exist in the absence of a complex neural substrate that includes a brain. There is no ground, therefore, to assume that psychic individuality arises during development before an adequate neural substrate has been laid down. But as later chapters will make clear, we cannot yet fully define a minimally adequate neural substrate. This leaves a considerable zone of uncertainty as to the precise time of onset of psychic individuality during the developmental course.

Becoming Social

Social individuality is conferred through recognition by others. No matter how strong the experience of inner self may be, it has no social content unless it is recognized by another. Recognition not only establishes societal membership but assigns a place and role in the social structure. So defined, social individuality is the opposite face of assigned status in society. It is not, therefore, an intrinsic property but a conferred one.

Societies differ in the stage at which they accord a full role and membership to offspring. Demonstration or acceptance of psychic (or other) aspects of individuality may or may not be prerequisite or judged to be relevant. Therefore, social individuality may be conferred, intentionally or inadvertently, either before or after psychic individuality is present.

However, this does not make the presence or absence of psychic individuality irrelevant in a society that is concerned about individual rights. In such a society, whatever the intrinsic state of the offspring, its perceived human quality may induce empathy—in turn leading observers to wish to assign a protective status. Therefore, when the developing offspring becomes recognizably human either in appearance or behavior, recognizability may itself become a significant factor in social reactions and decisions about its status.

This is illustrated by the emotional, and controversial, impact on observers of *The Silent Scream*. It also may be an important factor in the "burnout" reported to occur among health professionals who assist in numerous second-trimester abortions. Such emotional reactions to fetal appearance and behavior cannot be ignored in considering status for the unborn. What moves people, even if it is not rationally grounded, is not inconsequential.

Moreover, it has been argued that desensitization of such empathic responses might generally dull moral consciousness—leading to reduced sensitivity to helplessness in general, including to those who are mentally handicapped or terminally ill. While there is no certainty that this would occur, the possibility evokes caution in the thinking of many people.

Preliminary Conclusions

What preliminary conclusions can be drawn from this summary of major steps in the development of individuality? First, that the six aspects of individuality represent a progression in levels of complexity and that this complicates the assignment of status. Genetic individuality arises in a cell by virtue of the properties of a molecular component—DNA. It continues to be transmitted to successive generations of cells and is expressed during the course of development and beyond.

Human developmental individuality is a multicellular property that arises as the cell population gets larger, becomes cohesive, and undergoes differentiation—that is, begins to become specialized for different functions. At the same time, the population of cells shows its earliest integration into a single multicellular entity.

Functional individuality begins with formation of specialized organs and parts, including interactions among them to serve the requirements of a complex organism destined eventually to lead an independent life. For the first time, a new individual resembling its own species is recognizable.

Moreover, at this time behavior begins with primitive movements—the initiation of behavioral individuality. The fetus within the uterus is active in ways that are at least in part independent of the mother. Somewhat later (at sixteen to eighteen weeks), it is such activity, or quickening, that makes the mother aware of a "live" entity within her.

Independent behavior detectable by ultrasonic imaging as early as six or seven weeks raises the question of independent volition and possible sentience, properties associated with psychic individuality. Is movement fully diagnostic of the presence of inner fetal experience, or is it only a harbinger? The practical answer to this question lies at least in part in the response of the mother or other involved persons. Does the capacity to evoke empathy justify change of status, even if the movements are without awareness? At the very least, it can be argued, movement represents a *premonition* of social individuality—movement can initiate interactions in the social or quasi-social realm.

From all this, then, the first conclusion to be drawn is that human development is a translevel phenomenon. It begins with a new combination of hereditary molecules (DNA) in a single cell. It moves through a multicellular state to an organism of increasing functional complexity. It culminates in entities that behave, have inner feelings and experiences, and become interactive with others in social groupings. How does one relate this emergence through several levels of fundamental organization to the essentially social nature of status?

Second, the six aspects of individuality arise separately, gradually, with significant overlap in time, but roughly in the order in which they were discussed. If each were independently essential to full status, it would be reasonable to withhold such status until all six were clearly present to some defined level. On the other hand, if each is deserving of some increment of status, stepwise changes might be in order.

Third, a case might be made for considering the *set* of six to be more significant than their simple sum, perhaps because of the overlap and interaction among them. For example, genetic individuality is expressed in all subsequent aspects. Thus developmental individuality arises within a hereditary pattern provided by the evolutionary history of the species. So, too, functional individuality stems from developmental organogenesis and is a necessary base for behavioral and psychic individuality. In this sense, the character of the entity itself, in all of its aspects at a given stage, might be the indicator for status assignment.

Thus the concept of continuing emergence of individuality cannot be ignored in considering status for the unborn. The concept recognizes that the term unborn covers a period of enormous change by almost every criterion. Individuality in its several aspects comes successively into existence. Each new increment fundamentally changes the nature of the developing entity, in ways not needing consideration when assigning status to adults. These evolving aspects of individuality, separately and in their relationships, must be taken into account in any humane policy linking levels of individuality to status.

On the other hand, is an analysis of the relationship between individuality and status by itself sufficient to generate policy? What of the purposes and circumstances under which policy will operate? For example, regarding abortion, a proposed status must be tested not only against the level of individuality of the fetus but against the circumstances under which, and the purposes for which, a particular abortion procedure would be carried out. Similar consideration is required in analysis of fetal research or fetal surgery.

Having preliminarily examined some of the complexities of considering individuality in relation to status, I shall postpone to

Becoming an Individual

later chapters the even more complex problem of relating the two to circumstance and purpose. In the next chapter I concentrate on the emergence of integration, with emphasis on the role of the nervous system, particularly the brain. Without this particular emergence (that is, of the human brain) the whole issue of status of the unborn obviously would not arise.

CHAPTER 3

The Neural Substrate of Individuality

SIX ASPECTS of individuality were distinguished in chapter 2. Here we turn to a preliminary discussion of the nervous system, because its activities are so central to the entire phenomenon of human individuality. In fact, it cuts across all of its aspects. Thus neural development begins as one of the earliest expressions of the genome; the first visible neural rudiment appears almost immediately after the onset of developmental individuality. Continuing structural maturation of the neural rudiment is a prominent feature during the course of organogenesis, and functional neural maturation is overtly displayed in behavior when early movements are initiated at six to seven weeks, approaching the traditional embryo-fetal transition at eight weeks.

During the middle to late fetal stage (roughly twenty weeks to birth), neural maturation has progressed far enough to raise serious question as to whether psychic individuality may have appeared in the form of minimal sentience. This possibility heightens the empathy and concerns of those who fear that the fetus may experience pain and suffering. The debate over the social status of the fetus in this period becomes even more intense in connection with decisions about fetal surgery and the appropriate treatment of brain-defective or brain-damaged fetal infants (see chapter 6).

Thus the level of maturation and function of the nervous system becomes a central component of the status issue as fetal development approaches behavioral, psychic, and social individuality. This chapter offers a sketch of neural development and a preliminary appraisal of several of the major status questions to which neural development is particularly pertinent. The discussion anticipates the more detailed consideration of the next three chapters, which deal sequentially with the nature and status of preembryonic, embryonic, and fetal periods.

Behavior and Neural Integration

Perhaps more than any other human characteristic, behavior, in the broad sense of movement, action, and cognition, is an expression of the nervous system. This is because, though each individual is unitary in behavior, each is a collective of billions of component cells engaged in a multiplicity of activities and interactions. How does the enormous busyness of billions of cells get integrated into a single behaving entity?

Though it is not the only element involved in this integration, the nervous system makes two major contributions. Individual nerve cells, or neurons, provide rapid communication among other cells, tissues, and organs. Also, the neurons exchange signals among themselves in patterns of neuronal circuitry that have been organized in accord with earlier-established genetic and developmental individuality. The overall patterns, the complexity of which is only now being partially unraveled, underlie the unified behavior. *Thus the nervous system is the structural and functional substrate for the behavioral, psychic, and social individuality that become the substance of status.*

Neurons and Information Flow

Within the nervous system, individual neurons are the conduits along which informational signals flow. The informational signals move through a diverse population of billions of such neurons—most transmitting over distances far larger than the limited dimensions of ordinary cells. Neurons do the trick by forming extensions, or processes (see figure 16). In large organisms like ourselves, some of these cellular processes reach as far as several feet.

Information moves along the surfaces of the processes in volleys of short electrochemical pulses that can be recorded as abrupt changes, or "spikes," of electrical potential (see figure 13). By means of such specialized conduits, other cells or groups of cells can send or receive information, even over such very long distances as those from the brain to the tail of a whale. Large and complex organisms like humans and whales could not respond quickly and effectively without organized long-distance informational exchange of this kind.

Thus neuronal transmission is the essential substrate that enables large and complex multicellular organisms to function and behave as coordinated entities instead of as assemblages of self-serving cells. As a first approximation, the patterns of neuronal transmission can be likened to wired electrical circuits. The complexity of the circuits rises to the highest level—still only dimly understood—in our own brain. Here a hierarchy of communicating patterns forms a command summit that may be the most complex entity in the universe.

The Neural Substrate of Individuality

The Spinal Reflex Arc as a Prototype

The neural substrate for simple movements gives some conception of the possibilities and problems of applying objective scientific knowledge to the status of the unborn. I referred in chapter 2 to the neuronal circuit (reflex arc) associated with head-turning responses of the early fetus to tactile stimulation around the mouth. I will carry the matter a little further here, particularly in relation to the developmental history.

In adults, similar reflex arcs exist, though they rarely function in so simple a way (see figure 17). For example, special tactile sensors in the skin may respond to mechanical pressure by initiating a neural signal (volley of spikes) in a neuron. The neuron carries the signal *toward* the central nervous system (referred to as afferent). One end of the neuron is in the skin, the other end is in the spinal cord.

In the cord, the incoming neuron terminates on a second one lying entirely within the cord. This so-called interneuron is activated by the incoming signal and propagates it over its surface to its other end. Here the second neuron terminates in close contact with a third one, the motor neuron. It has a process that extends outward to make contact with a muscle cell, completing the neuronal sequence from stimulus to response.

Even the simplest actual behaviors have a more complex substrate than the idealized neural arc just described. Nonetheless, the principle holds that reflex behavior requires a series of neurons in sequence, from the periphery to the central nervous system and back to the periphery.

However, the idealized basic sequence is modifiable in several ways: (1) different senses (light, touch, sound, or other "modalities") may feed in to the same interneuron; (2) the several inputs may be either inhibitory or excitatory, and the response of the interneuron will vary accordingly; (3) different resultants are possible depending on the connection made by the motor neuron, for example, contraction of muscles, secretion by glands, changes of circulation, and so forth; and (4) the interneuron may make different or multiple connections within the nervous system, thereby distributing the incoming message to the opposite side of the body or to higher or lower levels along the spinal cord and into various parts of the brain (see figure 17).

The Synaptic Switches

Critical to these possible variations of neural processing of information are the junctions or synapses between successive neurons. At these junctions the two neurons are not continuous; rather their surfaces are pressed closely together without actually touching (see figure 12). These localized areas of close contact allow signals not only to be transmitted but to be controlled in their transmission. The resultant is integration of incoming information by the synapse.

The integration happens in the following way. Earlier it was mentioned that neurons transmit information in the form of electrochemical signals. The signals propagate rapidly along the neuronal surface, the intensity and duration of the stimulus being represented by the number of spikes and the interval between them. That is the maximum information that the signal itself can contain.

But synapses can add something more, because the electrochemical impulse is transformed to a chemical one as it crosses from one neuron to another. At each synapse, the arriving electrochemical impulse triggers the release of a chemical substance (called a neurotransmitter) into the narrow space between the two neurons (see figures 14 and 15). The transmitter diffuses to the opposite side of the space where it binds to a surface receptor specific to it and initiates a new impulse in the surface of the second neuron. Thus the electrochemical signal is transformed into a chemical one on one side of the synapse and back to an electrochemical one on the other side. What is the point of this seemingly unnecessary hocus-pocus?

As noted before, electrochemical signals are the same in all neurons, except for their number and frequency. But the neurotransmitters are not all the same. In particular, some *promote* initiation of a new electrochemical signal in the second neuron and others *inhibit* it. Since a second neuron has many synapses on its surface, involving neurons coming from different sources and producing either promoting (positive) or inhibiting (negative) neurotransmitters, the initiation of a new electrochemical signal becomes a complex resultant of all the arriving information (see figure 15).

This means that the interneuron integrates the information reaching it from many sources in "deciding" whether to transmit it. Synapses therefore function as components of neuronal computers. The actual response to a particular incoming stimulus is a composite of information from a number of sources. Rejecting or combining inputs makes for options in behavioral response. For example, a reaction to skin pressure may be different depending on whether simultaneous visual information identifies a friend or foe.

To generalize, synaptic connections among neurons are essential to neural function and resulting behavior. When the connections are not present, neither can occur.

Complexities of Circuitry

The sources of information arriving at a given interneuron are not limited to a single level of the body, known technically as a segment because nerves entering and leaving the spinal cord are arranged in regular sequence with respect to the segmented vertebral column (see figure 17). Interneurons of the adult spinal cord not only link incoming and outgoing messages within their own level or segment, they also send and receive messages between segments up and down the length of the cord (they are thus called intersegmental neurons).

This becomes apparent when severe spinal cord damage interrupts communication between segments or between a given segment and the brain. Such interruption can lead both to loss of sensation and to paralysis below the injury, even though local segmental reflexes remain intact.

The intersegmental neuronal processes run up and down the spinal cord in bundles of fibers known as tracts. Fibers within a tract have similar origin, function, and destination. The regularity of the arrangement provides patterning of communication that further integrates complex behavior (see figure 18). The origins or termination points of different tracts usually are collections of neuronal cell bodies, each of which is referred to as a nucleus.

Nuclei are sites of extensive synaptic connection, with inputs to and from other nuclei. Therefore, nuclei integrate incoming information at still higher levels. Moreover, some of these nuclei not only respond to inputs by providing outputs but spontaneously generate outputs of their own.

Sources of Circuitry

All of this complex circuitry makes up the essential neural substrate for functional and behavioral individuality, as we observe it in others or experience it in ourselves. The details of the circuitry are under intensive investigation, using ingenious and sophisticated tracer technologies. For example, electrochemical impulses can be traced from one end to the other using very fine electrodes inserted into neurons. Similarly, detectable "marker" substances can be injected into the cell body to follow its distribution out to the ends of even long processes.

But when and how does the complex circuitry thus revealed arise? What is the course of origin of neurons, synaptic connections, reflex arcs, and spinal cord tracts? In some fashion, the arrangements must stem, at least in part, from genetic individuality that is expressed in the course of realizing developmental individuality. If we know the times of origin of the neuronal patterns, the neurotransmitters, and the synapses, might this information be useful in assigning status?

To address these questions, we must return to the early stages of development of the nervous system. In chapter 2 I described the first appearance of an embryonic axis in the two-layered embryonic disc of the blastocyst. I said that a layer of cells involved in the emerging axis thickens into a plate that is the rudiment of the entire central nervous system (see figure 4), including the spinal cord and all parts of the brain. The term rudiment means that a precursor is identifiable, not that the actual system is mature and functioning. In fact, in the stage in question, neural maturation sufficient for function does not exist anywhere in the embryo.

A little later, during the third week after fertilization, the neural plate folds into a tube that separates from the originating embryonic surface. The larger forward end of the tube is now recognizable as the rudiment of the brain. However, at this stage the cells of the neural tube are not neurons. They have no processes and form no synapses. Rather they are growing and dividing, and they look very much the same whether they are in the region to become spinal cord or the various parts of the brain. They are as yet unspecialized or undifferentiated for neural function.

The rudimentary brain and spinal cord are thus structurally

continuous from head to tail from their very origin and remain so in the adult. Their linearity persists in the adult in the continuity of the neural canal, with its cerebrospinal fluid, within the spinal cord, and upward through the brain stem into the cerebral lobes of the brain. The integration that underlies functional and behavioral individuality depends heavily on this fundamental axial structural organization of the nervous system.

Becoming Truly Neural

How and when do actual neurons, synapses, and neural arcs first appear? Specific information based on human studies is not as abundant as one might wish, due to both technical and ethical constraints on the kind of studies that can be carried out. But when available information on humans is combined with more extensive knowledge about other species, the following picture emerges.

At about the fourth week of development, in the wall of the neural tube close to where the transition of spinal cord to lower brain will later be recognizable, some cells stop dividing and begin their maturation into neurons. Early identifiable neurons are present at about the fifth week, with neuronal processes beginning to extend toward the periphery (these are referred to as efferent, or motor, fibers).

A little later, during the sixth week, neurons that are forming processes can be observed in the wall of the spinal cord where sensory inputs will arrive (afferent, or sensory, fibers). Moreover, interneurons appear and the first synapses with them can be recognized. This suggests that the earliest continuous neuronal circuitry for reflex conduction and behavior could be initiated as early as six weeks. However, the number of synapses remains quite limited until the eighth week, when it increases noticeably, as it does again at approximately the twelfth week.

The Neural Substrate of Individuality

These observations correlate very well with the time of onset of observed movements, as noted in chapter 2. Based on both direct observation of exteriorized embryos following therapeutic abortion and more recent observations with ultrasound imaging, there is early movement at six to seven weeks, and the movements increase in diversity and strength thereafter. The overall interpretation is that the observed movements are expressive of a developing neural substrate and do not occur either prior to or independent of it.

Circuitry as Essential Substrate

These observations and conclusions come as no surprise to neurobiologists who accept as a first principle, based on a vast variety of information on many species, that behavior is the other face of the specific patterns of neural function that they observe. It is therefore to be expected that movement cannot occur until the necessary specific neural substrate has been laid down. Nonetheless, the unequivocal demonstration of the fact in the human embryo is consequential, since many people believe that human beings are very different from animals—if nowhere else, at least in the nervous system. And while indeed human beings are different from other animals in important ways, clearly they are not in the dependence of their movements on a suitable neural substrate.

If this conclusion can be generalized, certain projections from it seem reasonable. For example, it leads to a fair presumption that no human characteristics dependent on a nervous system are present during the first two weeks of development, when not even a rudiment of the nervous system exists. The same statement also would apply to the third and fourth weeks of development, when a neural rudiment is present but neither neurons nor synapses have yet formed.

These are not trivial conclusions, as will be evidenced in subsequent discussion. They suggest that if the exact time when a particular behavioral or psychic characteristic arises has not been established but its necessary neural substrate is known, the characteristic can reasonably be assumed not to be present if the neural substrate is not. The capability to draw such conclusions underlines the considerable importance of having available better and more detailed understanding of the development of the human nervous system.

Spontaneity versus Reflex

It is important also to emphasize that embryonic movements in the uterus appear to occur without external stimulation. There has been some question whether the ultrasound itself might activate the movements, but no supporting evidence for this possibility has been offered. Moreover, apparently similar spontaneous movement has been observed in other vertebrate embryos that normally develop externally and do not require ultrasound for observation. If such movements are not reflexive, in the sense of being direct responses to stimuli, how do they arise and what is their significance?

The point is worth discussion because a simple response to an external stimulus can be viewed as "caused" by the stimulus and not needing further explanation in terms of inner volition that might be akin to psyche. On the other hand, "spontaneous" movement seems more difficult to understand without invoking a source of inner activation, whether psychic or some other.

One possibility is reaction to an inner stimulus source, such as might arise in muscle itself. In adults, it is well known that proprioceptive reflexes maintain normal posture based on information received about the position of various parts of the body relative to gravity. Such reflexes do not require volition. In fact, the autonomic division of the nervous system constantly communicates information on alimentary, cardiovascular, and respiratory status, leading to responses that are not volitional and are largely free of awareness. Autonomic stimuli of this kind might figure in internally generated behaviors that are still fundamentally reflexive.

A second possibility is initiation of activity within the central nervous system itself, without any incitatory information from the periphery. Such spontaneous signal generators, which need no assumption of volition or consciousness, have been reported in a number of animal species. It also has been suggested that the reticular formation in the adult human brain stem may perform some such function, a point to which I will return later.

Pain and Recognition of Inner Experience

Our discussion has now reached two questions that are critical for many people concerned about status of the unborn—do they undergo some form of inner experience and, if they do, how can we recognize its earliest occurrence? For most moral positions, these questions are crucial because imposition of pain and suffering on the innocent is a cardinal desecration to be avoided at any cost.

But what do we know about whether the unborn can and do experience pain? As mentioned, the matter has been raised in connection with abortion. Pain is also a matter of considerable general medical concern since, in intractable form, it is a serious clinical problem in adults. The following discussion is largely based on reports of investigation of adult pain, emphasizing what appears to be relevant to the unborn.

First, adult pain is definitely dependent on a neural substrate.

The substrate, though not fully defined, is known to have several identified neural requirements.

Second, two aspects of pain have been distinguished. One is referred to as informational, meaning that it allows identification of the source and intensity of a stimulus emanating from a part of the body that is being damaged or is being treated in a way that threatens damage—for example, it is being subjected to heavy pressure. The second aspect of pain is the experience or feeling itself. The separability of the two becomes clear from the following information.

At a point of painful damage, a receptor initiates a neural signal that is transmitted into the central nervous system in the usual way for a sensory input. The distinction between the two aspects of pain is indicated when, in clinical observations, a patient makes appropriate movements to avoid continued noxious stimulation but, on questioning, reports no sensation of pain.

On the other hand, a patient can report the experience of pain without being able to localize correctly where it comes from. For example, pain may be only vaguely localized as abdominal, or it may be very precisely but falsely identified as to source—for example, in a limb that has already been amputated. And lesions in specific parts of the brain, or effects of anesthetic drugs locally applied to the brain surface, may suppress the experience of pain even though responsive movements are being made in an effort to avoid noxious stimulation.

Moreover, the experience of pain frequently is accompanied or quickly followed by more general feelings not engendered by informational pain alone. Fright, anxiety, and anger may be part of the general overtone evoked by experienced pain, but they are less likely to be evoked by purely informational pain.

For all of these reasons, it has come to be accepted scientifically that while the experience of pain overlaps knowing where it comes from, the two aspects are separable and presumably, therefore, are not identical in mechanism.

The Neural Substrate of Individuality

A third significant understanding about adult pain relates to its neural circuitry for transmission (see figures 17 and 18). The specific receptors for pain initiated in the skin are naked endings of unusually thin nerve fibers whose cell bodies are in the segmental ganglia ranged along the spinal cord just outside of the vertebrae. These same cell bodies also have processes extending into the cord to make multiple synapses on interneurons. Many of the interneurons, in turn, send processes into the spinothalamic tract to terminate in synapses on cells of the thalamus, the most forward region of the brain stem just below the cerebral lobes. The thalamus contains a complex array of nuclei (collections of neuronal cell bodies) and interconnecting fibers. It has long been regarded as a kind of gateway to the cerebral cortex, which in turn is believed to be the site of the most complex processing of neuronal information. And it is from the cortex—for example, through the downward-projecting corticospinal tract—that integrated information begins its return to lower segmental levels and the periphery.

What has just been described is a kind of cartoon of what is known about neural circuitry that includes interneurons in the brain, lying between an initial peripheral sensor for pain and a final effector also at the periphery. The fact that such circuits exist in adults and underlie informational pain is fully established. Behaviors that rely on such circuitry, therefore, *cannot occur until the necessary neural substrate has formed and is functional.*

Based on our current state of knowledge, conclusions beyond this must be much more guarded. For example, the notion of wirelike neural circuits is an analogy that is valid, in that it explains many aspects of observed neural function. But the analogy is subject to many limitations in the light of accumulating knowledge. It is becoming increasingly clear, for example, that on entering the thalamus via the spinothalamic tract, information is projected into what more resembles a complex three-dimensional network than a linear array. Moreover, it recently has become clear that neurotransmitters function not only at synapses but can be distributed through diffusion in tissues and even into circulatory channels. And this says nothing of the complexities of cortical information processing.

What Is Experienced and Where?

The consequences of spinothalamic input to the thalamus may, therefore, be less like transmission in linear circuitry and more like broad suffusion of areas of the brain stem. Under these circumstances, it is a matter of speculation as to how and where the inner experience of pain may emerge. The thalamus, the reticular formation, and the cerebral cortex are candidates for involvement, and all three may participate jointly. In relating this to the status of the unborn, the matter is not made easier by the fragmentary information about the development of these key areas in humans. The role of the brain stem in inner experience therefore remains an enigma, guarded not only by technical difficulties but by ethical ones.

What conclusions, if any, can be drawn about the general question of pain in the unborn? Since neither behavior nor its simplest neural substrate exist in the unborn prior to at least six weeks of development, there is no objective basis for assuming even the most minimal inner experience, including pain, during the first half of the first trimester.

The same cannot be said with equal certainty in the second half of the initial trimester, when new and increasingly more localized movements occur and the neural substrate is also increasing markedly in its maturation below the level of the brain. On the other hand, if inner experience and pain depend on a significant degree of brain function, they cannot be present at any time during the entire first trimester (thirteen weeks).

The Neural Substrate of Individuality

In fact, based on the limited information available, it can be said that brain-dependent inner experience such as pain is unlikely even up to twenty weeks. This follows from still-incomplete knowledge about the maturation of the human thalamus and cerebral cortex, two brain areas that appear to function interdependently and both of which are thought possibly or probably to be involved in inner experience. The first synapses among cortical neurons themselves do not appear until about eighteen to twenty weeks, and the first fibers from the thalamus do not enter the cortex until about twenty-two weeks.

Still later, at perhaps twenty-five weeks, thalamic fibers begin to synapse with cortical neurons and the latter begin to develop the extensive branches, studded with prickly spines for synaptic connections, that characterize the adult cortex. This process of vastly enhanced connectivity among cortical neurons presages a change in the electrical patterns observed in the brain via electroencephalograms. The patterns tend to become more regular and to show resemblance to adult patterns associated with sleeping and waking states. Such criteria lead some investigators to suggest that an adequate neural substrate for experienced pain does not exist until about the seventh month of pregnancy (thirty weeks), well into the period when prematurely born fetuses are viable with intensive life support.

By this late fetal stage, however, existing scientific knowledge is stretched quite thin as a base for fundamental and humane policy. In particular, knowledge of interactions among the thalamus, reticular formation, and cortex in relation to primitive kinds of inner experience is rudimentary and speculative. It is almost certainly premature, for example, to conclude definitively that no form of inner experience based in the brain stem is possible prior to cerebral maturation. Indeed, to find out whether this is the case should be a high-priority objective. It will require exceptionally creative approaches to yield insight without encroaching upon essential ethical constraints.

Conceiving the Minimal

Psychobiological research might also address two other questions: What would be the expected identifiable content of a minimal inner experience? And what would be the nature of its neural substrate?

Considering only the first question, how do we imagine a minimal inner experience? What is the *sine qua non* of subjective existence? Presumably it would consist of mere *awareness of being*, what has been called sentience. Such minimal sentience might be steady or fluctuating, either due to its own intrinsic mechanism or because of the nature and strength of the informational inputs to it. Such primitive awareness might function simply to heighten sensitivity to inputs and thereby to lower thresholds for response—that is, it might be a rudimentary stage of what in more mature form we would call arousal or attention.

To summarize, current knowledge of the nervous system establishes beyond scientific doubt that behavior in human beings has an essential neural substrate. It follows that, in the course of development, behavior cannot appear until an adequate neural substrate has been formed. Inner experience, which scientifically is viewed as resting upon and probably an extension of function and behavior, therefore also requires an appropriate neural substrate. This substrate almost certainly arises during maturation of the brain, particularly the upper brain stem and the cerebral cortex (see chapter 6).

Greater certainty and better judgment of the time inner experience begins requires deeper understanding of its neural requirements, especially with respect to the brain. This understanding can come only from a broad program of research, specifically motivated by the fundamental policy issues involved and conducted with full sensitivity to the ethical requirements that must be observed. The possibility that such research will be rewarding has increased with recent technological advances in detecting and measuring the activity of specific brain components in the living state. More about this will be said in later chapters.

CHAPTER 4

Policy for the Unborn: The Preembryo

IN THIS CHAPTER we begin discussion of policy issues that relate to each of the several major phases of unborn life, focusing first on the earliest and biologically simplest phase, the preembryo. This name itself deserves some comment because it has recently come into use amid some mild controversy.[1]

Disputes over terminology are not unusual in science, particularly in areas of rapid advance. What is unusual about this case is that a scientific term is called into question more because of its policy implications than on grounds of its scientific appropriateness. This is one among many examples of increasing interaction of scientific and social policy considerations in societal issues that are raised by advancing technologies.

Terminology and the Biological Nature of the Preembryo

In the last several decades, chiefly as the result of extensive studies of mouse development, it has become clear that in the earliest stages of each new generation, mammals (including humans) go through a preliminary preembryonic phase before they become embryos in the usual scientific sense. On reflection, this is not surprising, given the reproductive biology of this taxonomic group to which we belong.

The scientific concept of an embryo is one of an individual multicellular organism in the process of forming major parts and organs from rudimentary beginnings. In mammalian development, which normally occurs within the body of the mother (internal gestation), it is now evident that the early changes undergone by the zygote first establish multicellularity and, second, prepare for penetration into the maternal uterine wall, or implantation. The second step, as we have noted, is the true beginning of gestation or pregnancy.

In normal mammalian development, until implantation is achieved, there is not yet a rudimentary individual, or embryo. During the preceding stages, there is only an unorganized aggregate of precursor cells that lies within a cellular peripheral layer. The peripheral layer is busy with preparation for and then actual engagement in implantation.

Prior to implantation, the developing entity—although it is internal to the mother—actually floats freely in the fluids of her oviduct or uterus without direct contact or interaction with either. The physical separateness and relative independence from the mother during this period are made dramatically apparent after external human fertilization, when early human stages are cultured externally for a short time after egg and sperm are combined in a laboratory dish (referred to as *in vitro* fertilization, or IVF).

The earliest developmental steps in both mice and humans, whether *in vitro* or *in vivo* (in glass versus in the living organism), consist of a series of divisions of the original zygote cell. The absence of cell growth between these divisions distinguishes them from the usual cell division of later stages and leads to their being called cleavage, because with each division the size of the product cells is reduced roughly by half (see figure 1).

The result of the first several cleavage divisions is a cluster of smaller cells that are only loosely associated. In mice, it has been shown that each of the cells of the early cluster has essentially the same capability, in isolation, to produce a whole individual as the zygote cell itself. At least potentially, therefore, the initial cluster is precursor to several later individuals; in fact, in the armadillo the four-celled stage regularly gives rise to four identical offspring.

A little later in development, when the number of cells has been increased by continuing cleavage and the cells have become more closely compacted and adherent, most of the cells become segregated into an external layer, leaving a smaller inner cell mass projecting into a central cavity (see figure 1F). This is the blastocyst, the stage at which implantation into the uterine wall normally begins (see figure 1G).

The external cells, now designated trophoblast (feeding layer), are the elements that actively penetrate the uterine wall. They are precursor to the first cell population to differentiate or specialize. This population is the source of the embryonic component of the later placenta and of other extraembryonic structures that are significant in intrauterine life but are discarded at birth. The embryo itself will come entirely from the smaller inner cell mass, which, until implantation is well underway, shows minimal visible developmental change.

The earliness of the formation of trophoblast in mouse development was first strongly impressed on me personally some four decades ago when, working at the National Cancer Institute, I removed implanting mouse blastocysts from the uterus and transferred them to the eye chamber of adult mice. This seemingly bizarre maneuver was part of an effort to learn more about the properties of the early mouse embryo and how it forms. Others had reported that adult tumors and later-stage embryonic tissues were similar in their ability to continue to grow within the eye chamber, something they would not do in most other transplantation sites. To my amazement and consternation, instead of continued embryonic development when preembryonic stages were similarly inserted into the mouse eye, within a few days a hemorrhage occurred into the eye chamber.

It turned out that the developing trophoblast of the mouse preembryo was behaving as it normally would on contact with the uterine lining: It was eroding the surrounding tissue of the eye and opening blood vessels. If, prior to transfer to the eye, the outer trophoblast was carefully separated from the inner cell mass, the trophoblast alone produced a hemorrhage but the inner cell mass did not. Instead, the inner cells that are precursor to the embryo developed into a fascinating but poorly organized patchwork of growing and differentiating cells and tissues.

Such pathological development is referred to as a teratoma. Under some circumstances, it can become malignant—suggesting that there is some kind of inverse relationship between controlled normal development and the disorganized behavior of tumor cells.

The conclusion drawn today, on the basis of much additional information, is that the earliest stages of mammalian development do not primarily involve formation of the embryo or its parts but rather the establishment of nonembryonic trophoblast. This process is a necessary prelude to formation of the placenta that later provides the continuing essential exchange of substances between the embryo and the mother.

During the early stages, therefore, the developing entity is not best designated and understood as an embryo but rather as a preembryonic phase necessitated by the ancient mammalian commitment to internal gestation. The scientific rationale for the term preembryo, accordingly, is its greater accuracy in characterizing the initial phase of mammalian and human development. We will see, in fact, that this early phase differs from later ones in ways that are significant to assignment of status.

Anne McLaren, Director of the British Medical Research Council's Mammalian Development Unit, has told the story of her realization of this mammalian characteristic. While she was organizing a symposium on embryo formation in mammals, she recalls, "it first began to dawn on me that the 'embryo' as a continuous entity could be traced back from birth only as far as the primitive streak stage." As noted in chapter 2, this is the stage of the embryo during implantation and the first establishment of developmental individuality. McLaren continues, "It has taken a further ten years and some pressure from outside the scientific community for this distinction to result in a suggested change of terminology to eliminate the ambiguity of the term 'embryo.' "[2]

I had a similar experience at about the same time. While considering ethical aspects of IVF and other new reproductive options for the Ethics Committee of the American Fertility Society (AFS), I read reports of studies of early mammalian development after having been away from the field for decades. My assignment was to prepare a draft summary of early mammalian development as currently understood and relevant to the human species. Like McLaren, I was impressed by growing evidence that the early stages of mouse development were not embryonic in the usual sense but were dominated by the requirements of internal gestation. Considerable debate ensued within the AFS committee regarding terminology, leading to a consensus that use of the term preembryo would help to clarify the major watershed character of the developmental change involved in the initiation of an embryonic axis.

Whatever may have been the inner thoughts of the AFS committee members, the status implications of such a change in terminology were not at issue in the discussion. I fully agree with McLaren who, in responding to a suggestion that the term preembryo is a "cosmetic trick" (presumably to avoid constraints placed on "embryo research"), stated, "There is ambiguity in the way scientists use the term 'embryo'—and we are not justified in continuing to use the term embryo in both senses. *We are not talking about cosmetics but about clarity.*"[3]

Policy for the Unborn: The Preembryo

The Importance of the Status of the Preembryo

Why is status important in so early—and usually so inconspicuous—a phase of human existence as the preembryo? Actually, status of the preembryo was not a major focus of attention until several circumstances attracted notice. The first involves contraception, which is, of course, an intervention in human reproduction with the intention not to terminate pregnancy but to avoid its occurrence in the first place. The broad objective of contraception is to gain a measure of control over fecundity, whether narrowly to serve personal family planning or, more widely, to limit overall growth of human populations.

The contraceptive objective might be met by limiting sexual intercourse during the period of female fertility, a procedure acceptable, for example, to the Catholic Church, which is generally vigorously opposed to contraception. The objective might also be met either by blocking the meeting and/or fusion of sperm and egg or by interfering with implantation of the preembryo in the uterine wall. This second option has particularly raised the question of preembryo status, since in the current state of medical technology, to prevent implantation is to preclude development to birth and, hence, effectively to terminate the life of the potential offspring.

Opponents of this form of contraception, which is sometimes referred to as the "morning-after" approach, maintain that fertilization has already created a person with rights (status) and that to block implantation is to terminate an already created and sacred human life. In these terms, such contraception is tantamount to abortion and involves something akin to murder. This view was recently restated forcefully in the Vatican Instruction on Respect for Human Life (see chapter 1).[4]

The document asserts that the inviolability of the innocent human being's right to life "from the moment of conception to death" is both a sign and a requirement of the inviolability of the person to whom the Creator has given the gift of life. This teaching stems from "the light of Revelation." From the moment of conception, the life of every human being is to be respected in an absolute way; God alone is the Lord of life from its beginning until its end. Abortion and infanticide are abominable crimes.

Based largely on such Catholic and other fundamentalist religious teaching, a constitutional amendment has been sought in the United States (so far unsuccessfully) that would declare that a person, in the legal sense, comes into existence at the moment of conception. An amendment of this kind would reverse Supreme Court decisions in *Roe v. Wade* and other cases that have rejected the notion that preembryos have legal standing and rights.[5] In *Roe v. Wade* the Court dealt primarily with the privacy rights of the mother with respect to uses of her own body, denying any competing status to the unborn as a person until external "viability" at a much later stage of development. Therefore, the Court ruled on what the preembryo *is not* but avoided dealing with what it *is*.

Controversy over the status of the preembryo again erupted more recently in relation to external human fertilization. As noted in chapter 1, following fertilization human preembryos can continue to develop outside the body of the mother for at least several days. If cryopreserved (frozen by special procedures that minimize damage due to ice formation within cells), preembryos may be stored externally for weeks, months, or even years. The status and custodianship of frozen preembryos became a bitterly debated issue in Australia. A number of successful pregnancies have now been reported in many countries, including the United States, following transfer of frozen and then thawed preembryos.

John Robertson of the University of Texas School of Law has recently discussed the ethical and legal issues in cryopreservation of preembryos, noting that although "no jurisdiction has yet legislated directly on embryo freezing," official advisory committees in several countries have found the procedure acceptable in principle.[6] Nonetheless, it is the consensus of these bodies that substantial concerns about preembryo freezing remain unresolved and need to be addressed.

Among the concerns are the likely rate of induced abnormalities in resulting offspring; whether the preembryo has natural and/or assigned rights; the psychological and social impacts of the "time warp" produced during extended suspended animation; and the "slippery slope" that freezing clearly creates toward actualization of other forms of reproductive "engineering" in animals and, speculatively, in humans.

Can Science Help?

The question raised, in the context of issues discussed in this book, is what bearing, if any, scientifically established facts about preembryos have on issues of status. Do presumably objectively established characteristics of the preembryo speak to the issue of status and personhood?

Several preembryonic characteristics appear to be relevant. First, the preembryo is unquestionably human in biological terms. Certainly in common parlance, to be human is a minimal requirement for being accepted as a person. The preembryo meets the scientific definition of "human" in terms of such fundamental biological characteristics as the number, size, and shapes of its chromosomes; the sequence of the components (nucleotides) in its hereditary DNA; and the sequence of analogous components (amino acids) in its proteins.

But is biological humanness alone a sufficient condition to establish the presence and status of a person? The human characteristics listed for the preembryo are shared generally by human cells, tissues, and organs derived not only from living but also from dead persons. Such humanness is protected by both law and tradition against treatment that denigrates the dignity and value attributed to it by the human community. For example, there was general revulsion and condemnation when it was reported that the director of a Nazi concentration camp had lampshades made from the skin of executed prisoners.

Nonetheless, the question can be raised whether biological humanness is the same concept of humanness that has been promulgated repeatedly and vigorously in national and international declarations of human rights. Is the humanness of cells and tissues the same as that referred to in human rights declarations? Was the Declaration of Independence an assertion applicable to cells and tissues, or was it an expression of the indomitable spirit of people as political entities struggling to establish a right to be free from foreign domination? Can these two meanings of humanness, of cells and of people, be treated as if they are one and the same?

Second, preembryos awaiting transfer to the maternal uterus in an IVF laboratory dish are unquestionably *alive* by scientific criteria. They are exchanging respiratory gases, they undergo metabolism of their chemical constituents, their cells will continue to divide. They are, therefore, not only human but "in being." In fact, they are a stage in a human life history or cycle that runs continuously from a cell—the zygote—to a highly complex multicellular adult and back again to cells. The complex adult produces gametes that are specialized cells capable of fusion to combine genomes in the zygote, a cell with the very special characteristic of being capable of developing into another adult that can contribute new gametes. As has been said of poultry, a hen is only an egg's way of making another egg. And, throughout the entire life history, all stages are very much alive. If any were not, the life cycle would be broken.

Policy for the Unborn: The Preembryo

The cycle of life emphasizes yet a third characteristic of preembryos, probably the most distinctive and fundamental one with respect to formulation of their special status. It is a characteristic that goes well beyond those possessed by other human cells and tissues. Preembryos have a profound *potential*. They must be treated in ways that take into account not only what they are at the moment but also what they may become in the future—that is, an individual in the fullest sense, an undeniable person.

The potential of any stage of life (or, indeed, of any other phenomenon that is continuous but changing with time) is its range of possible futures. What happens to any stage in the present may change its future. In the case of human preembryos, potentially persons, this means that their treatment must take into account consequences for a future person, so long as the preembryo has a reasonable probability of continuing development to become an infant and then an adult. The situation is transformed if, for whatever reason, a particular preembryo has no reasonable prospect of developing further. In that instance, it would seem, a preembryo need only be assessed and valued for its then-existing properties without reference to what it might have become in a normal human life history.

Under these circumstances, the preembryo as it now exists is scientifically a human cellular aggregate, possessed of only minimal organization at any level above cells. So lowly organized a biological entity is vastly different in its nature and behavior from an infant and certainly from an adult. Its properties very much more resemble those of human cells or tissues separated from the body, though with an important difference still to be noted. In its entirety, a preembryo that will not, or cannot, continue normal development is still only a collection of living human cells. It has no evident *prima facie* claim to a status normally assigned to persons who are at the highest known levels of multicellular organization.

Yet certain human preembryonic *cells* are uniquely different from all other cells; because of this difference, they have a special value different from that of a person. Cells of the inner cell mass have the capability to divide many times and to specialize or differentiate into one or more of the plethora of cell types that make up the human body. Cells that have this capability have very high *potential as cells*, even when they will not produce a new and complex adult individual. This property makes them highly important to understand and, theoretically, makes them unusually valuable in practical terms.

I mentioned earlier that when inserted into the eye, mouse cells of the inner cell mass, comparable to those of human preembryos, can grow and differentiate into a teratoma, sometimes known to become malignant. Such preembryonic mouse cells have also been grown in laboratory culture, in some instances giving rise to lines of proliferating cells that still can differentiate into many kinds. Similar culture of preembryonic human cells might, therefore, open two possibilities: study of cell differentiation and malignancy in human cells and controlled production of specific tissue types useful in transplantation therapy.

Cell differentiation and malignancy are, of course, studied in cells and tissues from other organisms, as they properly should be. But if the time comes to apply what is learned on other species to human patients, tests on human cells in culture may well be judged to be essential before patients (persons) are put at risk in direct clinical trials. Cells and tissues derived from preembryos would be uniquely useful for these purposes, just as cells and tissues from adults are regularly used for similar purposes. If cells and tissues of individual adults, who are undeniably persons, can be so used under carefully monitored circumstances, why should the same not be true for preembryonic cells that will never become part of either an embryo or an adult?

Therapeutic transplantation of cultured preembryonic cells

and tissues is another possible benefit to be gained from the special characteristics of preembryos. Such therapy based on voluntary human donation is, of course, now widely practiced to replace defective cells, tissues, and organs (for example, blood cells, skin, kidneys) but is limited in its application both by available sources and by immunological incompatibility. At least theoretically these problems might be alleviated with cells and tissues derived from human preembryos. This possibility is being actively investigated with tissues from later stages, for the alleviation of the symptoms of Parkinson's disease.[7] Biologically, this would amount to maintaining preembryonic cells at the cell and tissue level of organization when the possibility of their development to higher levels of individuality has been excluded. This is a possible path to a cellular "fountain of youth" for therapeutic purposes.

Would such procedures denigrate the special quality of humanness? Clearly, this is where the matter becomes problematic. Some people have argued that use of preembryos for such purposes would reduce them to the status of implements or tools, not appropriate to the highest concepts of their human value. Others, I among them, believe that there is at least as persuasive a counterinterpretation.

For example, such use would give continuing meaning and significance, beyond simple respect, to human preembryos and their constituent cells. It would recognize in them value and status accruing from a uniquely consequential role. The role would stem from the preembryo's acknowledged biological membership in the human community, membership that need not be equivalent to that of a person but would merit more than the respect accorded to ordinary human cells and tissues.

The unique value is inherent in the special capability of cells that are precursor to the embryo to contribute to the welfare of other members of the human community. This unique role could be fulfilled without maintaining intact preembryos in culture beyond the fourteen-day limit recommended by several national deliberative bodies. The procedure might even be applicable to surviving cells of frozen and then thawed preembryos that had suffered sufficient damage to some of their cells to be unacceptable for uterine transfer.

The possibility of using preembryos in this way will certainly be troublesome to those who believe that respect for human entities, at any stage and in any condition, requires that they be regarded as ends in themselves and not as means to achieve the objectives of others, no matter how well intentioned. In this view, human existence is never a means to an end but always an end in itself. This is a deeply perplexing philosophical issue that goes well beyond simple concern for the future of the preembryo or its public policy status. It asserts that fertilization confers immediate moral status stemming from the nature and manner of the procreative process itself.

Here the values and orientations that are implicit in science are at some variance with particular philosophical doctrines and religious faiths. Current scientific knowledge says persuasively that the individuality created at fertilization is solely genetic— fundamentally important but limited to one among other important aspects yet to come. It can be argued that to confer full moral status on a genetic basis alone constricts and thereby denigrates the emergent profound nature of the developing human being, both as known to science and as recognized in ordinary discourse. This emergent nature includes, as noted in chapter 2, the genesis of developmental singleness, of functional complexity, of unitary behavior, of psychic identity, and of the capability for social interaction.

Thus in the specific case of human preembryonic cells, the inflexible religious view that assigns full moral status on genetic identity alone ignores the contribution to moral value of the rising cellular interdependence of the human phenomenon and especially the powerful role of interpersonal bonding and shared mutuality of interest. For primordial, preembryonic *cells* to become a symbol—and indeed truly a currency—for this profoundly human mutuality would seem to elevate rather than to denigrate their moral value and that of the preembryo of which they are part.

Analysis for Policy Purposes

The preceding discussion amply illustrates why the status of nascent human beings cannot be based on scientific considerations *alone*. Values, traditions, and religious teachings all strongly influence public attitudes. Moreover, with a population as diverse culturally as that of the United States, public policy can neither be expressive of one perspective nor can it realistically seek to incorporate all perspectives in a seamless blend. Rather it must attempt to identify and build on whatever general consensus exists, while remaining permissive and tolerant in areas of at least currently irreconcilable difference. It must also provide means of implementation that will support the consensus, without infringing more than necessary on the views and interests of those outside the consensus.

The role of science in this difficult process is to assure that the most reliable available knowledge is taken into account in the final outcome. How can this be accomplished in defining the status of the preembryo? I begin by distinguishing three categories of cases in which the question of status arises. Possible statements about status are then identified and interpreted for each category of case. Options for possible treatments and uses of preembryos are enumerated along with priorities among the options. Alternative decision-making mechanisms are then summarized, and, finally, attention is turned to specific issues that test a suggested status concept for the preembryo.

Questions about the status of preembryos have come up in three situations: (1) within the woman who produced the egg and will gestate the embryo and fetus; (2) external to the woman who produced the egg but with the expectation that it will be transferred to her uterus or to that of another woman for continued development; (3) external to a woman and not, for whatever reason, expected to continue development in accordance with the natural life history. The first category is the natural one, the second occurs regularly in IVF, and the third and currently less common category raises perhaps most acutely the question of preembryo status.

Four statements might be incorporated into a possible status for the preembryo. The first is negative: A preembryo is *not* a person. This is current legal interpretation in the United States but it remains controversial, with continuing efforts to modify it in practice and with the possibility that it may be overturned by altered interpretation in the Supreme Court. A substantial modification would have consequences for all three categories of cases.

The second statement is positive and also would affect all categories of cases: Preembryos should be respected for their biologically human quality. If the meaning of "human" were limited to its biological connotation—what is implied simply by membership in the species *Homo sapiens*—the statement would have the narrowest consequence. But if to be human is given moral content that is reinforced by political and other rights, the statement becomes both more consequential and controversial. And if "human" is to be regarded as equivalent to being a person, the statement obviously is the crux of the right-to-life battle. Any policy, then, must define what is meant by human in relation to the unborn.

A third statement goes beyond respect and is applicable only to the first and second cases. It says that preembryos must have a status that embodies awareness of, and special concern about, their potential to become persons—emphasizing that their treatment must never threaten or be deleterious to the welfare of persons-to-be. Such anticipatory concern, it should be noted, needs to be framed very carefully so that it does not become a virtual right of the preembryo, thus conferring *de facto* status of a person without declaring it. For example, this conferral of status already is occurring to some degree when IVF practitioners feel obliged to transfer to the uterus all available preembryos, instead of discarding or freezing those above the optimal number, because of concern about possible right-to-life violations.

Finally, a previously less discussed but possible fourth statement would assert that, while not persons, preembryos are nonetheless members of the human community with specifically recognized kinship and important community roles within it. Accordingly, they are entitled to contribute to activities of the community in specified ways appropriate to their nature. Such a statement would not often be applicable to the first category of cases but might be to the second and often would be to the third.

What options do these possible statements about preembryo status suggest for the custody and rearing of preembryos, if it is assumed that bringing up the child is the prerogative and responsibility of the natural genetic parents unless there are strong reasons to the contrary?

When genetic parentage is not possible, a second choice is adoption, either by relatives or by unrelated couples who want to raise a child. Thus adoption recognizes membership in the human community and gives priority to kinship when possible. Unlike conventional adoption, adoption of preembryos begins with gestation following transfer of a preembryo to the uterus of a woman who did not produce the egg and therefore is not a genetic parent. The frequency of such adoption will presumably be considerably increased if cryopreservation becomes more common.

If stable rearing arrangements cannot be made in one of these ways, other options include use for therapeutic transplantation and for research relating to high-priority medical or nonmedical objectives. Such uses would require specific sanction and oversight and would stem from the fourth statement just mentioned, establishing membership in the human community. They would signify that discard of preembryos under any circumstance should be only a last resort, one to be avoided if at all possible, given the profound value that attaches to all stages of humanity, whatever the stage of the life cycle.

Such possible uses of preembryos raise the question of decision authority—who makes the choice among the options? In natural reproduction the need for decision does not usually arise; about half of the zygotes terminate spontaneously and about half become preembryos that follow the ordained path to the uterus of the woman from whose ovary the egg was ovulated. She gestates it and traditionally she and her husband, as genetic parents, rear the offspring.

But reproductive technology has expanded the options, making it possible for ovulation, gestation, and rearing to be carried out by three different women, with or without voluntary male participation. Indeed, it has even been suggested (fancifully at the moment) that gestation might be carried out by men, and child rearing entirely by men is not a fanciful option. With these additional complexities, questions of parental or surrogate responsibility and of the assignment of dispositional authority for preembryos is a reality for parents, physicians, courts, and legislatures.

Among public groups that have examined the question, there is general consensus that first responsibility and authority should lie with the genetic parents, despite well-publicized but presumed to be relatively rare cases in which natural parents fall short in discharging their obligations.[8] When genetic or legally certified parents are not available, however, they may be replaced by various surrogates. These may involve: private trustees who are designated by the parents or by other private parties; public trustees who represent society; or public welfare agencies especially designated for the purpose. Although such questions of custody are not central matters of preembryo status, the list of decision options highlights the fact that the future potential of the preembryo is being considered and indicates that it is being thought of as, at least in part, a responsibility of its community.

Turning back from considerations of process to matters of substance, should the preembryo have the same protected status with respect to a right to life as does the newborn? The status of newborns in most contemporary societies is as strongly protective of continued life as is that of children or adults. Moreover, there are few proponents of legalized infanticide in western society despite its sanction or tolerance in some other societies.

On occasion, in fact, the strictness of the prohibition against infanticide in the United States produces severe moral dilemmas, as it does similarly in connection with marginal life situations in aged adults. This is especially true with severely defective newborns whose life expectancy is short and whose quality of life approaches zero. Particularly dramatic and tragic are so-called anencephalics with major brain deficiency. Setting aside these fortunately infrequent situations, should a preembryo have the same assurance of continued life as a newborn?

One principle that might be expected to be applied in comparing the status of two stages of development is the degree of similarity between them. If they are roughly similar, it might be expected that the two would have approximately the same status. In these terms, the huge difference between the preembryo and the newborn argues strongly against similar status. For example, the number of cells in a preembryo ranges up to several hundred, while in a newborn it is of the order of billions. Moreover, there are no more than two *kinds* of cells in the preembryo. In the newborn the number of cell kinds—depending on definition—is in the hundreds, and the diverse kinds are interlinked in a myriad of interactions that are both direct and indirect and occur through a variety of intermediaries that may be either other cells or substances produced by cells.

These interactions are so effective in creating the properties and unity of the multicellular entity that it is only in the last two centuries that it has been realized that the newborn, like an adult, actually is made up of billions of visually indiscernible cells. Indeed, taking these and other facts into account, a scientific judgment would have to be that there is far greater difference between the newborn and a preembryo than there is between a newborn and an adult. In fact, current evidence suggests that it took perhaps a billion years—about a quarter of the duration of the earth—for our ancestral prevertebrates to evolve from the simple cellular organization seen in the zygote and preembryo to the enormously more complex cellular organization of the newborn. Our far-distant evolutionary ancestors appear to have spent their whole lives as single cells or as small cellular aggregates, similar to bacteria and other microorganisms today. The state of the unicellular human zygote produced by fertilization was, in those distant days, the complete orbit of life for our ancestors, even unto maturity.

Having made the great leap to advanced multicellularity, humans nonetheless still return to a single-celled state in reproduction. The zygote cell, however, takes only nine months to again become the complex multicellular newborn. This is a tribute to the remarkable nature of development that should not obscure the enormous magnitude of the changes that occur.

Biologically, the preembryo and the newborn are so different —separated by nine months of development but a billion years of evolution—that it seems almost bizarre to think of them having the same status. To propose identity of status—whether moral, legal, or any other—seems not only to fly in the face of sound principles of logical classification but also to diminish the special quality that has evolved into fully developed humanity.

This is not to imply that the preembryo should not be assigned highly special value. Its unique potential should never be ignored or forgotten. But it also should not be ignored or forgotten that the preembryo is *yet unrealized* in its potential and that, until realization occurs, it is only barely emerging from the biological status of a simple cluster of human cells.

The conclusion, though difficult for some traditions to accept, is that the nature of a single human cell—marvelous though it is, especially a human zygote—cannot be confused scientifically with the very different entity that is the far more highly organized newborn. Most members of our society regard integrated multicellular human individuals, with the conventionally recognized characteristics of people, as persons deserving maximal human concern. A newborn has many but hardly all of these characteristics; the preembryo has essentially none.

Therefore, based on its scientifically established nature, the preembryo does not rationally *require* the same status as a newborn with respect, for example, to assured continuance of its yet quite limited individuality. Whatever the status of the preembryo should be, it should reflect the preembryo's existing characteristics and its potential without simplistically and arbitrarily assigning attributes characteristic of much later states.

It is worth emphasizing that even in natural development only about one in four human zygotes actually achieves development to birth. The high loss rate of three out of four zygotes is heavy in the preembryonic period, for reasons that are not fully understood. Genetic defects present in the egg and sperm and thereby incorporated into the zygote are believed to play a major role. These defects do not interfere with fertilization but exert harmful effects soon thereafter.

Whatever the mechanism of preembryo wastage, a particular zygote or preembryo clearly has no natural assurance of continued life; many fail even to implant. Of those that do implant, another significant fraction later succumbs to spontaneous abortion. As development progresses, the probability of continued life for the survivors rises. Thus, whether in the natural process or in artificial IVF, a particular zygote or preembryo has no guarantee of continued development. It has only the opportunity to play out its chances based on its hereditary message and its environmental circumstance. At the moment there is no known way to assure survival to every preembryo.

Preembryos and Research

Right-to-life questions are not the only ones that come up about the status of preembryos. As mentioned earlier, important questions also are raised about their appropriate custodianship, including who should speak for them in decisions about their fate. This problem emphasizes something central to the usual meaning of person. Persons are regarded as independent moral agents who must be consulted about their own welfare. But how does one consult a preembryo? Is it likely that there ever will be a time when it will be possible to do so?

Just to lay that question on the table is to make clear that preembryos are not persons in the usual sense of the term. They lack even rudimentary capability for the kind of communication usually relied on in identifying and dealing with persons. Who then should represent their interests—in whatever degree they can be said to have interests—parents, special counselors, physicians, institutions, society as a whole? As already described, this custody dilemma has been epitomized in troubling questions about the treatment of frozen preembryos.

But among the immediately pressing practical questions about the status of preembryos is their use in research, particularly to expand and increase the effectiveness of IVF. External human fertilization as a clinical approach to infertility had its origins in experimental laboratories studying animal reproduction.[9] It has now resulted in thousands of successful births around the world. But its success rate is only on the order of one in four to ten attempts, depending on how one defines success (for example, as the number of pregnancies achieved or of actual births). In either event, overall success rates remain significantly below that of the natural process and far below what would be desirable from the point of view of both patients and practitioners. Systematic research to improve the situation is limited by constraints and policy uncertainties that involve, among other considerations, legitimate ethical concerns.

A recent symposium on the problem was entitled "Human Embryo Research: Yes or No?"[10] It brought together almost thirty specialists in reproductive biology and medicine, genetics, law, ethical philosophy, social science, and technology. The group was challenged to consider the need for research on human preembryos and the circumstances under which it should or should not be done.

What is meant by "research" in regard to preembryos? Other words often used are observation, exploration, trial, investigation, and experimentation. Each has a slightly different connotation but all involve some degree of *systematic* study to increase understanding of a phenomenon of interest, thereby making it both more comprehensible and useful in practical terms. Simple observation is the least intrusive and penetrating of these approaches. At the other end of the spectrum, experimentation is probably the most intrusive, but it is also the most effective in penetrating to the nature of the phenomenon. In the case of IVF, for example, it would involve determination of the effect of various constituents of the culture medium on the developing preembryo.

The power of well-designed experimentation lies in its concentration on carefully defined questions and the setting up for comparison of controlled situations specifically designed to answer the questions. The effects of the composition of culture fluids on the development of human preembryos has never been studied experimentally because it is regarded as ethically and politically unacceptable to expose groups of preembryos to alternatives if one is presumptively more favorable than the other.

During the symposium, five medical areas were specified as in need of research on human preembryos: clinical diagnostic test for infertility, improvement of the efficacy of IVF procedures, improvement of contraceptive technology, enhanced capability for genetic diagnosis, and medical management of congenital malformation. On the other hand, these needs were set against the sensitivity and complexity of the ethical and social issues that are raised. The necessity to balance the two was recognized but no firm conclusions as to mechanism or policy were reached.

In the words of Sir Cecil Clothier, who provided an introduction and conclusion to the published symposium papers: "It seems shocking to reflect that most of the many sections of the Offences Against the Person Act of 1861 are still in force today [in Great Britain]. . . . This illustrates how far science has travelled while the law has almost stood still." In his concluding remarks, Sir Cecil noted that with respect to the five medical research areas, "there will be very little dispute that those are all worthy objects of research. . . ." But "none of these areas of research would be justified if what is proposed to be done is ethically unacceptable. . . . The central problem is the determination of the point . . . at which a group of cells is entitled to a special regard, or to some sort of rights." [11]

He concluded with the "uncomfortable feeling that the lawyers and philosophers have let the scientists down." On the other hand, the scientists "have presented us with the most frightful problems and confused us beyond belief. . . . I haven't grown any wiser but I feel much better able to discuss what should happen next and how and in what way we should formulate some rules which will both allay the proper anxieties of society and at the same time enable the search for the truth about human life to go forward properly."[12]

Research on preembryos in England is moving forward, although legislation to deal with the matter is still stalled in Parliament. In its absence, as a temporary measure, the British Medical Research Council, together with the Royal College of Obstetricians and Gynaecologists, established under their joint sponsorship a Voluntary Licensing Authority to oversee both clinical activity and research in human fertilization and embryology. The second report of this group summarized visits and responses to questionnaires sent to thirty approved centers.[13]

Fewer than half of the British IVF centers had research in progress, although all were attempting to improve the clinical efficacy of the procedures used. A number of centers reported current research relating to diagnosis of genetic defects and to improvement of the technical procedures being used, including cryopreservation. The work has been reported in sixty-four research papers and published communications. None of the research is in areas previously announced in Guidelines of the Licensing Authority to be nonapprovable, and all research had been submitted for review by local ethics committees. Under such voluntary auspices, the professional community involved in IVF in Great Britain is attempting to improve its services through research but is accepting and carefully observing constraints imposed by ethical limitations on human experimentation.

By Way of Summary

Preembryos are best regarded scientifically as a transitional stage between two major developmental landmarks—fertilization that establishes genetic individuality and the initial organization of a developmentally unitary individual. Fertilization is a cellular event; two cells fuse to become one—a zygote whose joint heredity is derived from both parents and that is activated to continue further development. Preembryonic development transforms a cell into a multicellular assembly, in preparation for initiating implantation into the uterine wall. It therefore chiefly involves early production of the peripheral trophoblast to mediate interaction with the uterus. In the natural course of events, embryo formation begins during the process of implantation.

The preembryo clearly merits the same respect accorded to all other human cells, tissues, and organs. This means that its significant human character—its kinship relationships as well as its genetic individuality—should be neither neglected nor denigrated. Beyond this general status, however, the preembryo requires additional special concern for its potential to become an infant and eventually a full person. However, if that potential no longer exists, for whatever reason, the value of the preembryo as a member of the human community should still be recognized and conserved.

This means, among other things, that the preembryo should be accessible to carefully considered research, providing that the research supports substantial medical objectives and the proposed plan for the research has been approved by a suitable review body. Failing achievement of its highest potential—to become a full person—the preembryo can thus realize part of its human heritage and potential by fulfilling a significant and unique role in the human family.

CHAPTER 5

Policy for the Unborn: The Embryo

THIS CHAPTER focuses on the status of the embryo. We saw in the last chapter that the preceding stage, the preembryo, ends when primary organization of the inner cell mass has produced a still quite rudimentary but nonetheless *single* entity— the early embryo. Its onset is conveniently marked by the appearance of the primitive streak. Visible with medium microscopic magnification, the streak is a transient, more opaque line on the embryonic disc that appears during the second week after fertilization. It foreshadows and marks the direction of what will shortly become the long axis of the embryo.

Soon thereafter, along the body axis ahead of the streak, body parts and organs of the embryo begin to appear (see figure 4). The process of emergence is known as organogenesis—that is, the origin of such organs as the heart and associated blood vessels; the central nervous system including the spinal cord and the brain; the digestive tract with its associated liver and pancreas; excretory kidney and associated ducts; reproductive organs; and external features such as limbs, head, and face.

Thus the period of the embryo is the period of genesis of form and function, reaching not to its final maturation but to the establishment of its broad outline. In some measure it is like a thunderhead building in a summer sky; orderly configuration appears to emerge out of nothingness. It begins with the primitive streak as the equivalent of little more than a puff of cloud. As the cloud billows upward in ever-increasing size and conformation, so too the primitive streak presages and seems to direct the emergence of multiplex human structure and form. Unlike a cloud, however, the emerging conformation of the embryo stabilizes as it becomes increasingly apparent in its pattern and character (see figure 5).

Yet the embryo, except for the heart and circulation, remains a quiet and unresponsive thing, in the sense of motion and behavior. Its changes are entirely along the arrow of time, the resultant of steady production of new cellular materials and emergence of structure and limited function. In this sense, the embryo is obviously alive; it is growing and transforming at a spectacular rate. But its changes are predominantly ones of growth, elaboration, and differentiation of its substance. As an entity, the embryo remains quiescent; it is living flesh with little indication of spirit.

Having manipulated and dissected hundreds of mouse embryos in the course of study of their development, I can testify that, in the language of the quick and the dead, the physical experience of dissecting an embryo is more reminiscent of autopsy than of surgery. The embryo is, of course, neither dead nor quick; it is probably best described as "prequick" in that in handling it one is aware of what it will become but also aware that it does not respond as if it were yet there. There is nothing of the resilience and reactability of more mature tissues. Nonetheless, I would feel much less comfortable in dissecting a human embryo than that of a mouse—although I expect that the two direct experiences would be very much the same.

Policy for the Unborn: The Embryo

Thus, during the period of embryonic organogenesis, the embryo can be characterized as totally passive and behaviorless. There is only minimal indication of the complexities of the full human individuality yet to come. The embryo is also totally dependent on the uterine environment supplied by its mother, often with little awareness on her part. But even as the embryo is supported and nurtured in the maternal environment, through its organogenesis its own capabilities for more independent function are steadily evolving. In the course of the subsequent fetal period, this will lead to increasing autonomy of function and, eventually, to capability for independent survival (albeit with external assistance) after severing the maternal connection —in other words, to viability.

From this thumbnail sketch, it can be seen that the embryo is—above all else—an entity in the process of extraordinarily rapid, fundamental, and complex change. Its cells are increasing in kind and, enormously, in number. Moreover, the cells are migrating, aggregating, and specializing (differentiating) as they form new structures and initiate new functions. In addition, as the new structures and functions are linking into interactive functional systems, they are increasingly coordinated and integrated as the essential substrate for *one* complete and complex individual. This is the course of increasing functional individuality.

A clear indication of the rising functional individuality and integrative maturation is provided as early as six or seven weeks after fertilization when occasional primitive movements are initiated as a harbinger of behavior. These are in the form of weak, almost flickering twitches of the head and neck that appear to be random and uncoordinated but certainly represent the onset of premonitory "behavior." This can be taken to be the marker for transition from embryo to fetus, what technically and traditionally was arbitrarily defined to occur at the end of the eighth week after fertilization.

The objective of this chapter is to relate questions of status to the rapid and fundamental transformations that occur during the embryonic period. Such status must accommodate not only the nature of the changes as seen scientifically but the practical issues that have become increasingly pressing as biomedical knowledge and reproductive medical practice have advanced in recent years.

A good example of the practical issues is the increasing concern about the sensitivity of organogenesis to sublethal teratogens, agents that can induce structural and functional defects in surviving infants. Exposure to alcohol, to other drugs, or to workplace toxins can exert such effects, sometimes even before a woman knows she is pregnant. The importance to be assigned to effective prenatal care thereby rises correspondingly—in step with the biological, emotional, and societal investment that is steadily growing as development progresses through the embryonic period.

Assignment of status to the embryo is complicated by a major practical reality—the interactive relationship between the embryo and the mother, a relationship that is so close, intimate, and mutually dependent that it can be called symbiotic and has often been compared with parasitism. In maternal terms, pregnancy begins with establishment of this relationship—that is, with implantation of the preembryo in the uterine wall. The mother, who is fully a person with already established social status, becomes a temporary biological host to her immature and dependent offspring, an entity of still-uncertain social status. The offspring itself, at the time the relationship is initiated, can claim only genetic individuality. Developmentally, it is not yet fixed as a single individual and does not exist as even a rudimentary embryo. Other aspects of individuality lie entirely in the future.

In implanting, the offspring becomes an active invader of the uterine wall in a fashion not completely unlike a malignant tumor. Moreover, the invader begins to capture bodily resources from the maternal tissues and circulation, much as a tumor or parasite might. Although only temporarily "parasitic," and usually fully welcomed and accepted by the mother, the invasion by the embryo is not always or in all respects envisioned as a blessing. Unwanted pregnancies lead to a very large number of induced abortions, well over a million a year in the United States alone.

Moreover, the price paid by the mother of even a wanted child—in physical, psychological, and social sacrifice—is always consequential and sometimes turns out even to be life-threatening. Thus the relationship between the unborn and its maternal host is reciprocal and intense but not always equally beneficial to the parties involved. Once pregnancy is established, however, the intimacy of the association is such that the interest of neither party can be served without significant consequence to the other.

Indeed, a central issue in considering the status of the unborn is when in pregnancy a woman should be considered to be *two* individuals rather than one. The matter recently came up, for example, in the California Supreme Court.[1] State law allows capital punishment for multiple murders, and the case being heard involved a man who shot his pregnant wife. The question is whether the unborn can be counted as an additional murder victim, clearly raising question as to the status of the unborn.

In this chapter and in this book, the focus is on the developing offspring and its status. However, the mother's reproductive privacy and rights are important "externalities" to be kept constantly in mind. The immediate question in this context is how the rapidly changing nature of the embryo should be factored into an inherently close, complex, and often problematic relationship between mother and offspring. The point is worth making because there is now a new context in which the status of the embryo may be raised. The conventional context is the more usual and natural case of the embryo developing within the uterus. The other may arise when the embryo is confronted outside the mother, either because it never implanted (arising from an egg fertilized externally) or because it was separated from the uterus by spontaneous or induced abortion. I begin with the first case, with the issues generated by growing capability to diagnose embryonic abnormality within the uterus.

Diagnostic Access to the Embryo

Two relatively recent technological advances have provided new means of diagnostic access to the human embryo in the uterus. Imaging using ultrasound allows visualization of the embryo and fetus in real time (as events occur) without surgical intrusion and with excellent resolution. Ultrasound visualization now also permits the guidance of instruments for taking tissue samples (biopsy) from the forming placenta near the end of the embryonic period. The tissue obtained is made up of chorionic villi (which are derivative of the trophoblast) produced by the embryo and therefore having its chromosomal and genetic constitution.

Ultrasound imaging thus allows embryos (and, later in pregnancy, fetuses) to be visualized without intruding into the abdominal cavity or the uterus itself and at a much earlier stage than amniocentesis, which samples cells in the fluids surrounding the fetus. Ultrasound can detect a number of anatomical abnormalities as well as the nature and pattern of movements. Chorionic villus assay additionally allows testing for genetic defect or metabolic insufficiency at the cell level. These diagnostic techniques are turning the unborn into a patient in its own right.

Unfortunately, with the current state of biomedical knowledge and technology, in most instances the diagnostic capability exceeds the therapeutic one; that is, defects revealed by these techniques, such as Down's or Lesch-Nyhan syndromes, usually cannot be satisfactorily treated within the uterus. However, if the defects are regarded as severe enough, abortion becomes a medical option. If abortion is to be performed, maternal safety is greater earlier than later in pregnancy, and the risk of psychological trauma is likely to be less as well.

It is just here that the abortion controversy rages most fiercely. How does one determine what defects are "severe enough" to warrant termination of embryonic life? Is it *ever* acceptable to terminate the life of an embryo? Is the embryo a person to be protected against violation of its human rights? Or is the embryo only a part (or parasite) of the mother, entirely subservient to her rights, particularly of reproductive privacy? And, if the status of the embryo were to be somewhere between these implied extremes, what decision process should be invoked before the life of a defective and unwanted embryo is terminated by abortion?

Prevention of Embryonic Deficiencies

However, abortion is not the only possible approach to embryonic deficiencies. Another is to prevent deficiencies before they occur. This approach also raises questions of status. Susceptibility to induced developmental abnormalities sharply increases with the beginning of embryo formation and continues through organogenesis. As mentioned in chapter 4, damage to earlier-stage preembryos is likely to lead to premature death rather than to defect. But damage to embryos is often localized in particular parts or organs, thereby producing localized defects in otherwise viable offspring. Examples include cleft palate, anencephaly, and foreshortened arms and legs.

Studies of the frequency and nature of embryonic defects in both humans and animals show that both the degree and kind of defect vary with the nature and intensity of the damaging agent (see figure 8). They also vary with the stage at which the damage occurs. Each organ or part has its own developmental schedule; sensitivity to damage rises and falls in different organs and parts at different times relative to overall development. Thus there are transient periods of heightened sensitivity, known as critical periods, in the various parts of the embryo. These periods occur most frequently in the second to eighth weeks of pregnancy when organogenesis is at its height.

Why does developmental sensitivity to damage raise the issue of embryo status? Because a number of fairly common exposures—to radiation, alcohol, certain drugs, cigarette smoke—clearly can contribute to abnormality arising during the embryonic period. Current public health policy with respect to this risk in the United States emphasizes education and persuasion as the appropriate means to achieve widespread and effective prenatal care of the unborn. But might suitable definition of embryonic status support and strengthen the effectiveness of this public health policy?

The current policy assumes that a pregnant woman is already maternal in motivation and attitude, in the sense of wishing to protect her offspring against harm, even at some sacrifice of self-interest. But it also assumes that coercion of an uncooperative woman on behalf of her offspring violates her reproductive privacy. Therefore, a pregnant alcoholic, or one addicted to heavy smoking or drugs, cannot be legally constrained or punished for behavior that places her embryo at risk. Nor, obviously, can her embryo be moved to other custody.

The question of whether more aggressive action is warranted was recently raised in a legal action taken in California against a woman whose son was born with defects and died some six weeks later. Reportedly the newborn had small amounts of amphetamine in his blood. The woman, according to police investigators, had been medically advised against drug use during her pregnancy. The district attorney, concluding that her failure to follow this advice led to the baby's illness and death, charged her with a misdemeanor—failure to provide adequate support to her child—which is a criminal infraction punishable by a fine and confinement of one year in jail.[2]

Several related cases in other jurisdictions involve drug use during pregnancy, but reportedly none led to criminal charges. The California action was applauded by the general counsel of the National Right to Life Committee, who was quoted as saying "Obviously if a born child had been intentionally injected with lethal drugs like this, you would fully expect action by the state."[3] The comment clearly assumes the equivalence of the unborn and born, which is one question at issue.

On the other hand, Ellen Goodman, a widely syndicated columnist, noted that "[this woman], having lost her child, was arrested for disobeying her doctor's orders. If this is a landmark case, then the land is easy to assay. At the bottom of this slope is a country where pregnant women must live by medical rules in the custody of the law."[4]

Less flamboyantly, a major conference of experts on alcoholism and birth defects went on record as opposed to criminal penalties for women who abused drugs during pregnancy. And a staff attorney of the Reproductive Freedom Project of the American Civil Liberties Union (ACLU), preparing an amicus curiae brief in the California case, said "we cannot allow the focus of [the] discrimination to be shifted to the rights of the fetus."[5] The issue thus raised, beyond equivalency of the rights of the unborn and born, is the status of the embryo in the uterus during the period of maximal sensitivity to developmental damage. In this case the plaintiff's request was denied on the ground that the cited California statute governing child support does not apply. The judge noted, however, that a state can enact a narrowly defined statute protecting a fetus by controlling the activities of a pregnant woman. State Senator Barry Keene observed that the existing statute was "never intended to prosecute pregnant women for failing to obtain medical care."[6] It remains to be seen whether specific statutes to this end will be sought.

Nonetheless, protection of the unborn against induced defect is likely to be pursued either by appeal of the judge's ruling or by specific statutes. Something less than criminal prosecution but more than voluntary education and persuasion—for example, civil liability or court order—may be the consequence, thereby contributing to further delineation of the status of the embryo.

It should be noted that the lifestyle of the mother is not the only source of influences that may affect the embryo. Increasing numbers of women in early pregnancy may be involuntarily exposed to toxic materials in their homes or particularly in their workplaces. The concern for future persons, outlined in the last chapter for preembryos, would seem to apply at least equally to embryos since they are more likely to survive as defective individuals. If reduction of the incidence of such defects is an important ethical and social objective, an appropriately clarified status for the embryo may be necessary to assure it. However, status clarification rather than severe punitive action against individual mothers would seem desirable.

Status of External Embryos

The status of embryos that are or may be external to the mother is also demanding increased attention. Such embryos may be produced by abortion, whether it be spontaneous or induced. Induced abortion usually severely damages if not largely disintegrates embryos. Such embryos are lost to possible constructive use, whether for study or as a source of cells and tissues for therapeutic transplantation. In addition, strong opposition to such possible use stems from religious, psychological, and ethical concerns about violation of the integrity of the individual.

Policy for the Unborn: The Embryo

It can be argued contrariwise, however, that it would be far more fitting, in terms of human respect and dignity, to put human embryos to use in ways that are fully reflective of their value as members of the human community. Under this concept, abortion procedures should be modified to conserve rather than to destroy the integrity and life of the abortus. The possibility raises issues of considerable complexity that will be discussed further later.

In the future, external human embryos also may become available through the continued development of preembryos maintained by technological life-support systems—that is, so-called ectogenesis. Research in this direction with laboratory animals has had some success but has not been applied to yield human embryos except, in extremely limited instances, in connection with IVF. Nonetheless, the prospect is widely mentioned in speculative scenarios for the future that have evoked largely negative reactions.[7]

Specifically, the matter has come up in considering how long IVF preembryos should be maintained in external culture prior to transfer. Early success in producing IVF zygotes provoked the question of whether they would develop normally if transferred back to the uterus. In a few recorded instances, preembryos were maintained in culture beyond the usual transfer time in order to indicate their probable normalcy if returned to the uterus.[8] But no data on how long such human preembryos might be maintained in culture under optimal conditions have been recorded.

However, the Ethics Advisory Board of the then-Department of Health, Education and Welfare, in reporting its deliberations on IVF policy in 1979, recommended that preembryos not be maintained in culture "beyond the stage normally associated with the completion of implantation (14 days after fertilization)."[9] This suggests that the board believed there to be some substantial change affecting status at that time. What change might justify limitation of external culture to preembryonic and preimplantational stages?

The Ethics Advisory Board did not provide a specific answer. A possible answer, however, lies in the appearance of developmental individuality during the course of implantation. The achievement of singleness, closely associated with the appearance of the early embryonic axis, might be regarded as warranting new status. But why should such altered status be expressed specifically as a limitation on the duration of continued culture?

It seems likely that the board was concerned about *research* use of developing human embryos in continued ectogenesis. With respect to "research for other purposes" than increasing the effectiveness of IVF, the board said that

potentially valuable information about reproductive biology, the etiology of birth defects, and other subjects may be revealed through research involving human *in vitro* fertilization, without embryo transfer, and unrelated to the safety and efficacy of procedures for overcoming infertility. The Board makes no judgment at this time regarding the ethical acceptability of such research nor does it speculate about what research might be sufficiently compelling to justify the use of human embryos. Instead, it notes that application for support of such research should be submitted to the Board for ethical review.[10]

The language of the report suggests that the board did not have a consensus on the touchy subject of research on human embryos and chose to postpone the general issue until it could examine specific cases. In fact, the board had little opportunity to do this since it was not reappointed and ceased to exist not long afterward. However, what seems clear is that the board recognized a difference between research on preembryos and on embryos. In relation to IVF, it saw the first as ethically acceptable. The second it saw as not related to IVF and distinctly more problematic without consideration of specific circumstances.

A recent report of the Ethics Committee of the American

Policy for the Unborn: The Embryo

Fertility Society is somewhat more specific in addressing the question of research on preembryos; it attempts to set criteria for acceptability and nonacceptability. For example, "Basic research on human pre-embryos should be considered only when no adequate substitute is acceptable and only to procure data that are likely to be of clinical importance." In addition, "The Committee concludes that it seems prudent at this time not to maintain human pre-embryos for research beyond the 14th day of postfertilization development."[11]

As a member of the Ethics Committee, I appended the following personal comment: "The report notes the arbitrariness of this limit [fourteen days] and mentions its relationship to the onset of the definitive embryo. However, it does not offer any other reason why status with respect to research should change at this time. It can be argued that the case for research beyond this time is at least as strong as prior to it."[12]

The case for research referred to in the comment can be expanded here as follows. Research on human subjects is in general recognized to be essential, both to understand the human phenomenon and to improve human prospects.[13] Such research raises important issues as to purposes, limitations, and appropriate regulation. Procedures applicable to adults have been formulated, are in practice, and meet these requirements reasonably well. Particular groups of potential human subjects raise special problems: for example, children, prisoners, the mentally incompetent. These problems have also been carefully examined and are managed reasonably effectively. Why should the same not be done for embryos?

The answer that is usually given includes reference to the entitlement of embryos to continued life, to their innocence, their helplessness, the possibility of their experiencing pain and suffering, and their inability to provide informed consent. While these are all nontrivial limitations and problems, they do not appear to be inherently insurmountable. Nor are they all unique to embryos and fetuses.

In the transition from preembryo to embryo, the essential change is the establishment of the early embryo as the rudiment of a single complete individual. That change may well warrant a corresponding change of status, but why should it require what amounts to a *de facto* total prohibition against using embryos as subjects for research? If the preembryo, the infant, and all other stages of human life can, under specified circumstances, be subjects for research, why should the embryo (and fetus) be excluded? Rather than questioning whether research should be done at all, should the appropriate question be under what circumstances it might occur?

Obviously, the issue generates deep division of opinion. The recent Vatican Instruction referred to earlier declares that "all research, even when limited to the simple observation of the embryo, would become illicit were it to involve risk to the embryo's physical integrity or life. . . ." The only exception is "experimental forms of therapy used for the benefit of the embryo itself in a final attempt to save its life." Destruction of human embryos for research purposes "usurps the place of God" because the researcher "arbitrarily chooses whom he will allow to live and whom he will send to death."[14]

On the other hand, in a recent review of the legal, ethical, and policy aspects of embryo research, John Robertson notes that "Embryo research is essential for the development and refinement of a wide range of infertility treatments, as well as for understanding many other disease and developmental processes. Clear thinking about the issues raised is essential to formulate an appropriate public policy." He concludes that there is an "overwhelming consensus" among official and professional bodies that have reviewed the matter that "much embryo research is ethically acceptable and should be permitted."[15]

Under what circumstances should research on human embryos be sanctionable? Responsible deliberative bodies that have discussed the issue have already made some suggestions: *only* if there is a high-priority medical objective that cannot be achieved in any other way; *only* if a formal proposal has been evaluated and approved by a group, independent of the proposers, concerned specifically with the purposes and procedures of human experimentation; and *only* if there can be no detrimental effect on an offspring that comes into being. These provisos, if they were to be respected and enforced, would add up to assurance that research on human embryos would not be carried out routinely or casually and would not lead to socially unacceptable objectives.

Despite expressed fears to the contrary, there is no reason why research on human embryos *must* lead to complete external development (ectogenesis), human hatcheries, or other speculative scenarios of the Brave New World. In fact, there are few, if any, serious current research purposes for which culture of intact human embryos through their developmental course is attractive or necessary. This is not to say that serious motivating rationales may not by generated in the future (see chapter 8). But of greater contemporary biomedical interest are human cells, tissues, and possibly organ rudiments. These are far more readily maintained under laboratory conditions than whole embryos, and since they are devoid of potential to become a person, they are therefore less limited (but not entirely unlimited) by ethical considerations. And their practical medical applicability is considerable.

An already mentioned example is therapeutic transplantation. Other possibilities include vaccine development, susceptibility to cancer initiation, and factors involved in various congenital defects. Each area would have to be examined carefully in terms of necessary ethical and social constraints, with continuing oversight provided by mechanisms to be discussed in chapter 7.

Criteria for Status Change

Are there aspects of the rising individuality of the developing embryo that particularly call for a change of status? As form and function are gradually elaborated, two new qualities stand out in social terms: recognizability of external human features and bodily movement. Both have impact on observers and are keys to social interaction. Each is the product of complex internal change that registers in equally complex and subtle ways.

For example, the configurational changes that register as human facial features (see figure 6) arise out of cellular behavior—proliferation, interaction, migration, and production and shaping of intercellular materials. These activities increase overall mass and distribute it as external topography. As a result of these morphogenetic (form-generating) activities, major external features come into existence—nose, ears, eyelids, cheeks. These and other features gradually resolve into a tiny and still-incomplete human miniature.

At some point, varying to some degree with different observers, identifiable humanness generates feelings of empathy. These feelings are heightened by behavioral movements, particularly expressive facial behavior that appears to be initiated from within and to represent coordinated and intentional behavior. It is interesting to ponder the prevalence of both natural and artificial facial markings in both humans (including cosmetics and jeweled ornaments) and many animal species. Do these not function as signals to evoke empathic or even submissive responses?

As such maturation of signals of humanness advances toward some threshold, a seemingly reactive and responsive human being is increasingly perceived. This psychological effect on observers cannot be ignored in discussing status, not only because of the empathy engendered but because it is a forerunner of social recognition and interaction between the developing offspring and others. Thus the embryo, now a miniature human being in its major structural features and approaching capability for rudimentary behavior, imperceptibly undergoes transition into a fetus.

Arbitrary though the exact transition point may be, a distinction between embryo and fetus is important to maintain. During the embryo stage overall, major organogenesis occurs, whereas the fetal stage is characterized primarily by growth and maturation. The boundary between these two major periods of development has been difficult to establish accurately from traditional clinical data. As noted earlier, it is now more significantly, directly, and conveniently defined by the onset of movement as visualized by ultrasound imaging.

This suggestion fits the fact that the recent clinical availability of ultrasound imaging has provided new indices for staging human development in general. Moreover, movement as a criterion for staging is in accord with long tradition. Quickening, the first fetal movements detected by the mother as midpregnancy approaches, was relied on for centuries as a signal of the impending arrival of a healthy new offspring. However, the significance of movement in relation to functional maturation and possible sentience is a complex matter to be discussed further in chapter 6.

Should Embryos Be "Used"?

I have already referred to the strong moral tradition that regards human beings as ends in themselves, never to be treated as means to other ends. Whether that tradition is applicable to embryos depends on the status assigned to them.

In chapter 4, a distinction was made between preembryos that might or would become a person and those for which that possibility was precluded. The same distinction seems applicable to the embryo, but with the modification that, without external intervention, embryos fully implanted in the uterus are quite likely to continue to term, while intact embryos external to the uterus currently cannot continue indefinite development.

The question of alternate uses of embryos therefore arises particularly in relation to those that are external to the uterus, either as the result of abortion or of continuing ectogenesis from earlier stages. I have already referred to their possible usefulness as a source of cells and tissues for transplantation therapy. In theory, the intact embryo need only have developed sufficiently to yield a *rudiment* of the desired tissue or organ, since such isolated rudiments can be cultured and will continue their own development in isolation for considerable periods. In fact, animal studies indicate that in laboratory culture, isolated organ rudiments—for example, of lung or kidney—develop more fully than they would if within an intact cultured embryo. Similarly, isolated immature skin can grow to considerable quantity in laboratory culture and can be successfully used for skin grafting.

Should a human embryo, genetically and developmentally an individual but never, for whatever reason, to go on to higher levels of individuality, be used to serve the medical needs of other members of the human community? Such a role is, of course, similar to that of voluntary adult donors of blood, tissues, and even organs—with general approval. Is a similar role for an otherwise terminal embryo a violation or a strengthening of the concept of respect for humanness? In the broadest assessment of overall human benefit, is there a net gain or loss in thus treating human embryos? And if there is or may be a net gain, how and by whom should the determination be made in particular cases?

Therapeutic transplantation from human embryos raises still further issues. Assuming that such procedures would be successful, desirable, and ethically acceptable, would the number of human embryos derived as a by-product of other objectives be able to meet the demand? Or should human embryos be produced specifically for the purpose?

Were the last possibility to be sanctioned, the logistical bottleneck clearly would be the availability of mature human eggs. Should surplus eggs obtained in IVF procedures be cultured for this purpose? Should women be encouraged to act as donors, comparable to sperm donors, considering the greater risks entailed in harvesting eggs? Should such eggs be fertilized using donor sperm, without intention of their becoming offspring? If so, should this be an enterprise that could be pursued commercially? Or, alternatively or jointly, should research be encouraged on the production of eggs from ovaries in culture, perhaps starting with ovarian rudiments obtained from embryos?

If such technologies were as successful as the natural egg-producing mechanism represented by the queen bee, relatively few ovaries might suffice to produce ample numbers of embryos (now we *are* talking human hatcheries). The procedure, incidentally, theoretically could make possible an abbreviated human life cycle from ovary to embryo to ovary, with considerable speeding up of intergenerational genetic change, which would offer a substantial boost toward influencing human heredity over generations.

Such scenarios *do* suggest a Brave New World and *are* fanciful for short-term policy making. But they may not be fanciful in the longer term and in future contexts (see chapter 8). Rather, they emphasize the importance of realistically facing the widening range of possible interactions with the embryo, of understanding its real nature, and of carefully defining its status. The last is already essential to give policy guidance to health professionals who are on the front line of contemporary reproductive clinical research and effort. They need a defined standard of social acceptability that sets limits but does not unduly restrict legitimate possibilities for present and future benefit.

Clearly the integrity and future welfare of the embryo should be strictly protected so long as it may possibly become a complete individual. But, failing that possibility, the embryo may be most respected and given most meaning by permitting it those roles for which it is uniquely qualified and which fully express its essential unity with the entire human family.

Should Embryos Be Selected?

Prenatal diagnosis of genetic defects is adding new stress to these issues because, given a dire prognosis without possible cure, therapeutic abortion becomes an alternative to a possible long course of suffering and tragedy. In turn, for many people, decision on abortion practices rests on a credible and widely accepted definition of the status of the embryo.

Over three thousand human genetic diseases have been catalogued. Included are sickle-cell anemia, Down's syndrome, and autoimmune deficiency. The ability to diagnose genetic diseases is currently advancing far more rapidly than the ability to treat them. Although direct gene transfer may be able to correct some genetic defects (a possibility that is now being explored), current and the most likely near-term major application of advancing genetic knowledge is to early diagnosis by molecular techniques. If and when a severe genetic defect is reliably established and cannot otherwise be treated, competently performed abortion is a reasonable option for many people.

How severe and how certain must the effect of the defect be to warrant abortion? On one side is the view that *no* genetic defect is severe enough to warrant intervention if the embryo can possibly survive, regardless of the expected duration of survival or the quality of life. At the opposite extreme is the view that any defect is sufficient if it leads a pregnant woman to decide against continuing the pregnancy. But many people, perhaps the majority, subscribe to neither extreme and would want more information on the nature and severity of the defect, the likelihood that there will be early advance in available means of treatment, and whether and how much society is willing to help in meeting the burdens of extraordinary care.

Case-by-case assessment of these situations is laborious and expensive in time and emotional stress. But in early stages of innovation with respect to genetic medicine, such efforts may be preferable to oversimplified generalizations derived from traditions established in earlier and less well-informed times. Clarifying the embryo's overall status could give helpful guidance to those making the difficult and sensitive case-by-case judgments.

Should Embryos Be Shielded?

A third problem area that would be affected by clarification of embryo status has already been referred to—developmental abnormalities. At least some of these are known to be induced by toxic external agents, and as many as several percent of newborns have one or another abnormality. Perhaps the most striking example affecting U.S. opinion was the thalidomide tragedy in Europe some years ago. This drug, prescribed to reduce anxiety and sleeplessness in pregnant women, turned out to induce marked and characteristic limb abnormalities in their offspring. Even after thalidomide was identified as causal, its mechanism of action was unclear and its study was hampered because it did not produce the same effect in common laboratory animals.

The incident called attention to the fact that embryos in the uterus are not as cloistered as sometimes thought, and it contributed to the decision to shift March of Dimes support to congenital abnormalities after the development of the polio vaccine. It also emphasized that the results of drug testing in nonhuman species are not always reliable in forecasting human effects. This led to interest in possible testing of suspect materials, in critical cases as outlined in chapter 7, on cultured human embryonic tissues and organs in laboratory culture.

There can be no doubt that information derived from such testing could be useful and that the need is growing. The expansion of occupational roles for women is increasing the possibility of their exposure, and accordingly exposure of the offspring they may be carrying, to agents of unsuspected toxicity. As noted earlier, the embryonic period, even before a woman may be aware of her pregnancy, is particularly susceptible to induced abnormality.

This matter, like resort to abortion for genetic defect, cannot be addressed through overly simplified generalization. The number of suspect toxic substances in workplaces (or even homes) is large, and probably not all have been identified. Certainly not all have been tested for their capability to induce developmental defects. Valid testing of even a single one would require a substantial number of embryos. Testing a thousand would create a significant supply problem. If human embryos are to be used at all for testing, particularly critical cases will have to be carefully selected. The central point here is that the use of human embryos for such purposes should not occur before a suitable definition of embryo status is formulated and a credible decision mechanism for its application is in place.

Policy for the Unborn: The Embryo

Overall Perspective

The human embryo, from the time of establishment of its developmental individuality at roughly two weeks after fertilization, to the initiation of movement a little prior to eight weeks, changes rapidly and radically in its properties. Its status is currently unclear and subject to controversy. The embryo transforms from an initially quite undifferentiated multicellular state through establishment of complex cellular, tissue, and organ systems that begin to be integrated into recognizable structure, function, and behavior.

It is essential that the status of the human embryo during this rapid transformation be defined if we are to deal effectively with advancing knowledge that is steadily raising new issues, both in the present and for the future. No single biological event or process provides a clear and simple basis for distinctions of status during the embryonic period. Rather status must take into account the total complex that is emerging—an embryo that is alive, human, has kinship, and is prospectively a person, yet is short of this full realization in fundamental respects. In fact, the embryo can be characterized scientifically as still functionally incomplete, showing no behavior in the usual sense, and remaining so immature in its neural maturation that there can be no scientific foundation for assuming even the most minimal aspect of an inner life.

Nonetheless, biologically the embryo is clearly a member of the human family, and it begins to be recognizable as such as it progresses into the fetal stage. Its biological humanness, therefore, should be affirmed in the status assigned to it. But it also should be recognized that the embryo is still importantly short of the full individuality it will eventually attain.

Accordingly, the integrity and future of the embryo should be fully protected so long as complete development is possible. But when the integrity of the embryo is no longer meaningful because it cannot or will not develop further, it is still a member of the human community and should be valued for the special contributions it can make to that community. Its treatment and status should be defined in ways that are fully sensitive to its transitional role in the human life history. I shall say more about this in chapter 7.

CHAPTER 6

Policy for the Unborn: The Fetus

THE FETAL PERIOD of development is the most perplexing and complex to deal with in terms of status. The difficulty stems from the fact that the fetus is steadily approaching a boundary—the beginning of life beyond the womb, the termination of being unborn. In preparation, the fetus shows mounting functional and behavioral individuality. As the fetal period advances, uncertainty also rises as to the possible existence of minimal psychic individuality as well.

Since many people regard behavioral and psychic individuality as key to the assignment of status (and our factual understanding about neither is complete or fully persuasive), it is not surprising that considerable controversy surrounds fetal status. Moreover, there is yet much to learn about the fetus, especially about its level of neural maturation and brain function. This chapter, therefore, deals with an inherently problematic and contentious subject.

Looking at dictionary definitions, the controversy and complexity are not apparent. According to *Webster's Third New*

International Dictionary, the fetus is simply an unborn or unhatched young vertebrate that has passed through its earliest developmental stages and attained the basic structural plan of its kind. The human fetus, as I have noted, is traditionally said to begin at about the end of the eighth postfertilization week when the embryo, as previously defined, passes without sharp boundary into the fetal stage. The fetal stage then normally ends at birth when the newborn infant, by definition, comes into existence.

The dictionary does not say that *most* complex organisms go through intermediate states between the embryo and the adult —states that have other names than fetus. Thus butterfly eggs become creeping and feeding caterpillars before they metamorphose into reproductive and flying adults. Frogs have an aquatic stage as tadpoles that feed on vegetation. After a period of growth the tadpole metamorphoses into an amphibious, or even terrestrial, frog with an entirely different habit of life. The adult frog is carnivorous and reproductive, laying and fertilizing external eggs to start its life cycle over again.

In primitive mammals—for example, marsupial kangaroos— the fertilized eggs develop into embryos in the uterus, and these then emerge as early fetuses. The fetuses are mature and neuromuscularly agile enough to climb into the external maternal pouch where they fasten to a teat for continued nurture and maturation.

Though not a caterpillar or larva, the human fetus is also an intermediate stage in structure and function between the embryo undergoing basic organogenesis and the succeeding infant increasingly capable of limited independence of the mother. Like the intermediate stages of other organisms, the fetus is rapidly growing, and it is also steadily maturing to higher levels of function and behavior. Its increasing maturation is equipping it, for example, to change from an aqueous environment to a gaseous one in which it must become air-breathing and capable of exchanging respiratory gases in its lungs. At birth, and even before with special support, the normal fetus accordingly can survive without a placental link to its mother, although it still requires much nurture and care for a considerable time after birth.

A live-born fetus is termed an infant. In reality, a late fetus and an early infant are not very different either in appearance or degree of dependency on nurture and care. Fetal "viability" refers only to capability to survive disconnection from the placenta. It does not imply complete independence, which is not, of course, achieved until many years after birth. Moreover, the exact time and degree of viability after disconnection from the placenta is heavily influenced by the amount and nature of available supportive care, whether from the parents, a surrogate, or a specially equipped technological nursery.

The concept of viability and its relation to fetal status, as is well known, have become particularly controversial with respect to abortion. In effect, under current legal interpretation, viability confers new status on the yet-unborn fetus when it is known to be capable of survival externally with appropriate care. Prior to this capability, the life of the fetus is secondary to the mother's right to reproductive privacy, which includes the right to terminate pregnancy even though it means terminating the life of the fetus. After a fetus becomes viable, however, the state may assert a compelling interest to intervene. Thus the boundary rests on fetal viability, a boundary that can shift if improved technology for life support at earlier stages becomes available. The range of the mother's privacy right would thereby be constricted.

Pressures for Clarification of Fetal Status

Pressure to clarify fetal status also comes from directions other than abortion—for example, from the rising effectiveness of prenatal diagnosis and therapy. In recent years the fetus has come to be considered a second patient, rather than a secondary complicating circumstance in the care of the mother. Techniques like amniocentesis and chorionic villus assay, undertaken with consent of the mother, provide cells whose analysis can detect genetic and metabolic disease in the fetus, requiring new decisions that bear on fetal, maternal, and familial welfare. Should fetal interest have any voice in these decisions if it appears to be at cross-purposes with maternal and familial wishes?

Moreover, ultrasound imaging usefully visualizes both later-stage embryos and early-stage fetuses within the uterus. Defects in individual organs or organ systems can be detected. The technique of fetoscopy with concentrated light sources (fiber optics) also allows direct inspection of the fetus and even permits collection of blood from fetal vessels. The battery of techniques can yield a significant forecast of important aspects of the future of the fetus, thus offering choices as to course of treatment that may not have equal consequences for the fetus and other interests. How, then, is fetal interest to be considered and protected?

In many instances, unfortunately, the diagnosed problem may have no known therapy. Should the information then be sought in the first place? If the information becomes available and is dire, it forces a hard choice on the parents: whether to attempt to raise a severely handicapped child, with possibly limited life quality, or to abort it. Should the status of the fetus permit its life to be terminated under such circumstances, or should an alternative rearing arrangement be sought? If the latter, whose is the decision and by what procedure and criteria should it be made?

In other cases, potential therapies exist that need not interrupt the pregnancy—but they entail some risk to, and some cooperation from, the mother. The stern reality of the fetal-maternal unit then must be recognized. To treat the fetus, one must invade the mother, and her risk and rights must be added to the complexity of the medical decision. Given the interpenetrating intimacy of pregnancy, should the decision serve the interest of the fetal patient even though its mother is placed at significant risk? And if the medical decision is that it should be, what efforts should be made, if any, to persuade, pressure, or coerce an unwilling mother?

In all of these situations, fetal status is clearly one item in a difficult equation. A dramatic case in point was recently provided by treatment of a fetal kidney disease that involves blockage of the ureter draining the kidney. This abnormality can be recognized by ultrasound and is life-threatening to the fetus. Moreover, treatment often cannot be postponed to birth without jeopardizing the newborn irreparably. Following ethical review of several such recent cases, informed consent was obtained from the mother for surgical entry into her abdomen and uterus. The fetus was carefully exteriorized without disturbing the placenta, and the fluid was tapped from the affected kidney. The fetus was then returned to the uterus without inducing premature labor. In one particularly successful case, operated on at twenty-three weeks of pregnancy, the baby survived birth nine weeks later—when it was in the safe zone for intensive care of prematurely born infants.[1]

Technical capability for such diagnosis and direct treatment of the unborn is growing. The cited case is thus a bellwether for other therapeutic interventions on behalf of offspring in the uterus. For example, another recent report tells of surgery in the uterus on twins whose circulation was interconnected in ways that threatened the life of one and inhibited development of the other. In previously recorded cases of this type, both twins died prior to birth. In this case, surgery removed the twin most at risk and permitted the other to be born prematurely but alive.[2] Such therapy clearly requires weighing of fetal interests and status as well as maternal ones. Therefore, fetal status must necessarily be more carefully defined.

There is yet another reason why fetal status and interests are becoming increasingly pressing issues. Development of innovative medical procedures, illustrated by the cases just cited, is necessarily preceded by a period of clinical investigation and trial. Studies of animals are important preliminaries, but they alone cannot guarantee either the efficacy or safety of an innovation in actual practice. Sooner or later there must be testing on actual fetuses, at least initially without certainty of success. Yet current federal policy and regulations are so restrictive that they substantially inhibit clinical trials and thereby the advance of fetal medicine. Clearly the need is to define the status of the fetus appropriately, so as to give it the protection it legitimately needs as a human subject while not undermining the welfare of another fetus needing yet-unproved intervention.

Prematurity and Status

The central objective, then, is a fully and truly humane definition of fetal status. What status should the rapidly emerging but still-transitional individuality of the fetus have? At what point, for example, should the status of the fetus equal that of the neonate? In considering these questions, the issues arising around the management of prematurity are especially pertinent.

At how early a gestational age, given the costs and risks, should heroic measures be undertaken to maintain life in prematurely delivered but otherwise normal and healthy fetuses? Should heroic measures be undertaken if the fetus is substantially abnormal? In extreme abnormality are even "ordinary" measures necessary or desirable? Do any circumstances justify deliberate termination of the life of a live-born fetus? How should these thorny dilemmas be considered and resolved in crisis situations that require quick decision?

In the past two decades, these questions have become almost

routine for neonatologists, a group of subspecialists of pediatrics that concentrates on medical problems of the soon-to-be-born and the newly born. The subspecialty arose out of strong motivations of pediatricians and others to rescue prematurely born infants who were regularly dying in respiratory distress.

The lungs of these tiny newborns (often born at less than half of normal birth weight) are simply not sufficiently mature to allow adequate respiratory exchange in an extrauterine environment. Contemporary intensive care nurseries have come into existence since midcentury, when it was found that prematurely born infants often survived when maintained in an environment controlled for temperature and composition of respiratory gases (that is, in an incubator).

While such early beneficial effects of incubation were encouraging, they did not by any means overcome all problems. Many premature infants still died or failed to develop normally. More sophisticated respiratory support, using positive pressure to avoid lung collapse, and improved techniques for feeding and controlling infection further increased survival rates.

These advances led to a new lifestyle for some newborns, featuring pipes, tubes, and mechanical pumps instead of pink and blue blankets and dolls dancing on a string. The new life-support technology was applied in segregated pediatric intensive care nurseries presided over by specially trained new pediatric subspecialists in neonatology. As the technology advanced, it became routine to have premature infants survive after only thirty weeks of natural pregnancy, and even infants with gestational ages of only twenty-four to twenty-six weeks have survived.

However, even this happy course of medical advance was not untrammeled. Although many prematurely born fetuses survive in the special technological environment created for them, the costs are very heavy. Cumulative dollar costs rise rapidly with degree of prematurity because the less mature the fetus is, the longer the time it must spend in highly expensive intensive care. Time spent in the uterus is cheaper and, in many ways, safer. But there are also noneconomic costs, including increased risk to the health of the fetus, immediately and in the future. Although most treated premature infants survive and become normal children, a significant number do not.

As the number of such cases mounted, especially in fetuses born prior to thirty weeks of pregnancy, questions arose as to the wisdom of initiating intensive care in the first place. Are such decisions always warranted? Is "the game worth the candle" if it only lengthens the period of dying or provides a minimal life mostly of pain and suffering? But can one *not* initiate care if the fetus is born alive? And once initiated, by what indication and by what mechanism can one stop?

Conference on a Dilemma

A report of a conference on such questions appeared in 1975 in the professional journal *Pediatrics*.[3] The conference examined in detail the cases of five infants born prematurely at twenty-eight to thirty-two weeks of gestation. The infants were given full technological support in an effort to assure their survival. Two of the cases produced young children, normal at the time of the report. The other three cases produced children with severe abnormalities that limited both the duration and quality of their lives.

Analysis of the five cases led to formulation of three clinical questions: (1) Is it ever right not to resuscitate an infant at birth? (2) Is it ever right to withdraw life support from an infant clearly diagnosed as having a poor prognosis? (3) Is it ever right to intervene directly to kill a dying infant?

The conferees included "twenty persons . . . from different

disciplines: medicine, nursing, law, sociology, psychology, ethics, economics, social work, anthropology and the news media."[4] The diverse group was unanimous in its opinion that there could be circumstances (not unanimously specified) when it would be right not only not to resuscitate but to withdraw life support from an infant clearly diagnosed to have a poor prognosis. Seventeen of the conferees believed that certain circumstances could justify killing a slowly dying premature infant. Two dissented from this opinion and one was uncertain.

Another Crucial Status Boundary?

However representative the opinion of such a group, the basic issue raised—now frequently and starkly faced by physicians, parents, and others—is the status of the fetus at about thirty weeks when it is in transition from middle to late fetal stages. Should the status of the late fetus include an absolute claim to continued life comparable to the entitlement already extended to the infant? Should the premature late fetus and the normal-term infant be identical in this aspect of status? And if the late fetus that has been born should have such entitlement, should the unborn late fetus have it as well? Or should birth itself confer the status?

The dilemma expressed in these questions arises from the fact that while an infant born at normal term has a high probability of survival with normal care, a prematurely born infant has a distinctly lower probability of survival unless it receives special care. The greater the prematurity the lower the probability of survival without special care. But the greater the duration and intensity of special care, the greater the monetary and emotional cost and the greater the risk of severely damaged survivors. Given this tortured calculus, it is not surprising that the parties to decision seek guidance in more formal definition of fetal status and the role of the parties involved.

Anthony Rostain, a physician exposed to these clinical dilemmas, recently provided an insightful analysis of the decision process with respect to some of these questions, particularly the possibility of involuntary euthanasia before the age of informed consent when life prognosis is extremely poor. He describes the decision process in intensive care nurseries "like it is," with many persons playing difficult and tense roles in an atmosphere of high uncertainty and considerable emotional stress. He suggests need for and likelihood of "new social mechanisms [that] will be developed to rationalize the decision-making process."[5] I shall return to this basic issue in chapter 7.

The difficult dilemma is intensified by right-to-life insistence that fetal life be continued, whenever and however possible, as an absolute value in any circumstance. This allows no distinction between one hour and one year of continued existence, nor between an hour of suffering and an hour of well-being. Moreover, it offers little comfort or help to parents and caretakers already caught up in a dreadful dilemma.

In some premature spontaneous deliveries, parents and health professionals are confronted by a fetus that is demonstrably alive but that best professional judgment says will not survive many days or, if surviving longer, has no likelihood of awareness of its own existence. Under such circumstances, right-to-life insistence seems not to rest on the value of life but on the desperate need to ward off death. Understandable though this may be in emotional terms, it is difficult to incorporate into a rational definition of fetal status—since, sooner or later, every individual must face the reality of death.

On the other hand, a fetus is clearly well along in its development of individuality, and this must be taken into account in considering its status. A preembryo is inconspicuous enough to be flushed unnoticed down a drain, a late fetus is not. The already extended period of pregnancy has involved investment of maternal and other time, energy, and emotion that heightens value and commands preservation. This value, however, must be judged in the light of all considerations and interests, not on a single absolute standard rigidly adhered to.

To be more specific, the status of the fetus beyond roughly thirty weeks must take into account that it is already essentially an infant if suitable external support is available. This, however, is at a cost, both in effort and in somewhat greater risk to the offspring's well-being. If the life of the resulting newborn is projected to be normal, the status of the prematurely born fetus need be no different from that of a normal newborn. But if the projection is for a severely limited life—a burden not only on itself but on its family and society—a special situation is created that needs sensitive definition of status and special mechanisms for decision (see chapter 7).

The Interests of the Future and the Unusual

What other considerations and interests are relevant to decisions about fetal status? Certainly future potential, as noted earlier for the unborn in general. A significant part of the original potential has now been realized, and the fetus should be valued accordingly. But not all potential is yet realized, and treatment of the fetus must continue to show concern for future impacts of current decisions.

This should include the expectations of the offspring itself at a later age as well as the expectations of parents and kin who will be significant in later life. Beyond immediate family is the larger community that may derive benefits or incur costs. Such aggregate evaluation is complex and difficult and not routinely necessary. But in special cases it may be essential to a wise decision, and provision must be made for it.

Involved parties, of course, view the individual offspring within their own set of concerns and interests. In uncomplicated pregnancies, involved parties other than genetic parents—chiefly the health professionals who advise and consult with them—briefly play defined roles and move out of the scene.

But matters are not always routine, as is the case, for example, with anencephaly, a tragic condition referred to earlier. It is among the most severe congenital anomalies that still permit live birth. Its origins lie at the beginning of organogenesis: due to improper formation and closure of the neural tube, most of the brain may be missing while the brain stem is relatively intact. In some instances, this permits development to birth, and survival for a short period beyond, because circulation and respiration can be maintained by brain-stem centers.

If there is electrical activity in the brain stem, an anencephalic newborn cannot be ruled to be dead under the now widely legislated definition of death designed for adults—total absence of electrical activity in the brain. Nonetheless, if not stillborn, every anencephalic dies within a few days of birth. The general question has been raised whether such marginally living fetuses may provide a source of tissues and organs for therapeutic transplantation.

The possibility arises because successful transplants, particularly of the heart, previously have saved the lives of some infants born with serious cardiac defect. But infant donors are so rare that even the heart of a baboon was once used in a desperate attempt to save the life of a defective human infant.[6] The much-criticized attempt in California failed because of immune rejection, but legislative efforts have followed there to allow transfer of the heart of an anencephalic, since this heart might be expected to be normal despite the absence of the brain. (In a recently reported California case, an anencephalic's heart was transferred successfully to a defective newborn. The anence-phalic donor was delivered in Canada and, with parental and ethics committee assent, was maintained on artificial life support during shipment to California.)[7]

To accomplish this objective legally might require legislation to declare an anencephalic to be the equivalent of brain-dead. Law professor Alex Capron has recently discussed the legal complications of such a maneuver,[8] and ethicists Frank A. Chervenak and associates have analyzed the ethical status of anencephalics with special emphasis on abortion during the third trimester (beyond twenty-six weeks).[9] The ethical analysis was undertaken for that late period because anencephaly may not be diagnosed until then. The analysis suggests that abortion is morally justifiable if two conditions are satisfied: (1) the donor fetus will not survive more than a few weeks after birth or, if it does, it will have no cognitive function; and (2) the diagnostic procedures that provide the forecast are highly reliable. Chervenak and associates feel that anencephaly meets both conditions. Based on their search of the literature, they believe that no other defect currently meets the criteria, though "a small number of additional defects" may do so in the future.

Their ethical argument is based on the principle of benefi-cence, "to do no harm to the fetus and to provide benefit to it whenever possible." Since the anencephalic is fated to die with-out cognitive function, it can be argued that neither harm nor good can be done to it. On the other hand, termination of the pregnancy provides a good to the parents by shortening the time to a more productive pregnancy and by eliminating the psycho-logical trauma of continuing a futile one.

Clearly, this argument is analogous to cost-benefit analysis and will not necessarily be acceptable in all ethical belief sys-tems. However, it also fits the policy concept of gradual ac-quirement of protected status during the fetal period, with cir-cumstance playing a significant role on a case-by-case basis.

An interesting twist on the cited ethical principle is provided by a recent case in Georgia.[10] A woman carrying a twenty-one-week fetus was hospitalized for drug overdose and pronounced brain-dead. Her husband asked the hospital to withdraw life support but another man, claiming to be the father of the unborn and not-yet-viable fetus, sought a court order to continue support at least until the fetus was viable externally. The court accepted the plea against the state's assertion that the court had no jurisdiction at least until the fetus was actually viable.

In this instance, although viability usually limits the maternal right to abort, it was invoked to assert the right of a previable fetus to continue its life—through support to a neurologically and legally dead mother. Whether the ruling establishes a precedent for earlier status of the fetus is not clear. In effect, it turns the dead mother into a component of a life-support system and, in this special circumstance, pushes viability of the fetus back to twenty-one weeks.

There is another side to this increasing concern with preserving the life of the fetus. Ethicist George Annas sees "the beginning of an alliance between physicians and the state to force pregnant women to follow medical advice for the sake of their fetuses."[11] He cites a study by V. E. B. Kolder and associates reported in the same issue of the journal.[12] As the result of a questionnaire sent to directors of training programs in maternal-fetal medicine, these authors found twenty-one cases in which court orders were sought on behalf of fetuses for obstetrical intervention against the wishes of pregnant women. The procedures involved caesarean sections (a procedure requiring entry into the abdominal cavity and uterus to deliver a fetus), forced hospital detention, and intrauterine transfusions. In more than 80 percent of the cases court orders were successfully obtained within six hours after the request. The authors also found that approximately half of the training directors queried favored detention of mothers whose actions threatened the health of their fetuses.

The authors are alarmed by the facts they report, seeing "serious ramifications" for the welfare of minority women who predominated among the cases identified. They believe that the question raised is "whether doctors or the government may usurp patients' decision-making rights and appropriate or invade their bodies to advance what they perceive to be therapeutic interests of a second patient, the fetus."[13]

Here the anticipated community of interest among physician, pregnant woman, fetus, and state seems to fall hopelessly apart. Some new policy guidance is required to protect the fundamental right of the woman to avoid intrusive invasion of her body while not losing sight of the growing claim of the fetus to protection as well.

Birth as a Status Marker

The concept of a premature infant implicitly assumes birth to be a primary status marker—since "premature" means less mature than at normal term. Moreover, birth is a widely accepted status marker (though not a universal one) in societies generally. In our society birth is essential, for example, to establish inheritance rights under common law. Nonetheless, some commentators have suggested that birth is overemphasized as a landmark event. They argue, quite correctly, that birth represents no sharp change in fetal characteristics and, as demonstrated dramatically by prematurity, may occur at very different levels of fetal maturation. Why, they ask, should it involve so discontinuous a change of status?

The counterargument, however, seems more persuasive to most people. Birth certainly is a landmark event in the life of parents. And for the infant, even if it is not recalled, it is a watershed because the neonate is now exposed to an environment far different and far more varied than that of the uterus.

Perhaps most important, however, birth initiates direct interaction between the infant and others—it is the onset of the mutual social behaviors that are the substance of status. For these reasons, it is unlikely that natural birth, as a profound translocation from a primordial to a lifetime environment, will ever be ignored as a landmark in human life history. Whenever and however it takes place, even if it were merely "decantation" into a Brave New World, it thrusts the infant into a fundamentally new physiological and social existence that influences all that follows. To be newborn is very different from being unborn.

This leaves to be resolved the difficult issue of the status of the late fetus prior to birth. If birth changes status primarily because it changes relationships, does that mean that the intrinsic nature of the fetus has nothing to do with the matter at all? On its own merits, should the late fetus have different status from the infant?

Several kinds of rationale can be applied to the question. Answers can come from authoritative teaching, religious or otherwise. They can come from frankly empirical and political negotiation. They can stem from collective formulation of explicit objectives and their priorities, followed by a marshaling of the best available knowledge to realize the objectives. Any enduring policy that will be widely supported will have, in some measure, to combine these rationales. The following discussion emphasizes accommodation of rationales, giving priority to objectives that can be both clearly articulated and evaluated by reliable facts.

For many and, very likely, most people, the crucial issue of fetal status is the existence or nonexistence of psychic individuality—that is, the question of sentience and consciousness. With little agreement as to the exact nature of the phenomenon, we all identify it in the primary experience of self. We call this experienced self-reality *I*, a continuing *one* that is distinct and bounded not only from other selves but from all else that is nonself.

Because each such self-identity is anxious and concerned about its own state and fate, each is also empathetically anxious about the welfare of other selves, particularly helpless and vulnerable ones. There—but for the grace of God—go I is a common theistic formulation. More generally, if I were being treated as that fetus is being treated, how would I feel? Then the question easily follows, when does something like my inner feeling of self first occur in the course of development? In scientific terminology, what is the course of the genesis of self (egogenesis)?

The Problem of Self-Genesis

The problem is more than a little troublesome. It is fundamentally problematic despite centuries of philosophical wrestling around the edges. With all of its advance, contemporary science claims no certain answer though it begins to see avenues along which the question might be approached.

For example, one conceptual approach might be to specify the minimal characteristics of what we call self in terms of characteristics that might be objectively identifiable. Two characteristics appear to be essential: awareness of simple existence and awareness of a boundary between existent awareness and all beyond. In other words, self is an inner sense of existence that excludes whatever awareness there is of other things around and beyond. For our current purposes, this is an adequate definition of self—minimal psychic individuality.

Under this definition, what would be left if either awareness of existence, or of the boundary to it, were to be deleted? Deletion of the first leaves nothing to bound, so the two disappear together. But deletion of the second leaves an unbounded, diffuse locus of awareness that seemingly fills a space of indefinite extent. Such an imaginable phenomenon might be a predecessor to what we call self-awareness. Experiences of this general kind have been described as occurring in people who have suffered severe brain damage or are in drug-induced altered brain states. Such awareness without boundary, designated diffuse sentience, might be close to a rudimentary form of self. If one could establish, in adults, associated objective signs of such a state, one might be a step closer to identifying the initiation of self in the fetus. In defining status of the fetus, one might then feel more secure in avoiding traumatic intrusion into minimal diffuse sentience.

When might this kind of diffuse sentience arise in the fetus? We have already seen (in chapter 3) that current knowledge supplies no certain answer to this question. No direct experience from the fetal period is reliably recalled by adults or even young children; few if any adults remember anything earlier than the second to third year after birth, presumably not because there is no experience at this early time but because recall is not ordinarily elicited in interpretable form.

There is helpful indirect information, though on many points it is not conclusive. It comes from several sources: levels of maturation of the brain in human fetuses, similar information on other species, and behavioral studies on both human and animal species. This indirect information offers several tentative but relevant conclusions:

1. During the fetal period new objective characteristics relevant to sentience are steadily arising. They include elaborate maturation of the central nervous system ranging from molecular and electrical properties to complex neurological circuitry.
2. This massive neural maturation allows division of the fetal period into early, middle, and late subperiods for status purposes. The early fetal period ends at about twenty weeks, the middle period at about thirty weeks, and the late period at term (forty weeks).
3. The early fetal period begins with primitive movements based

on simple neural arcs, accompanied by maturation of external features to levels that begin to generate minimal empathy. Functionally, however, the early fetus is still quite immature, particularly at the level of the brain.

4. The middle fetal period centers on the rise of viability, meaning that fundamental physiological functions are maturing toward independence from maternal support via the placenta. Still particularly limiting are the immaturity of the skin and the respiratory, immune, and nervous systems. At the beginning of the period these deficiencies cannot yet be compensated for by current external life-support technology; by the end of the period they can be.

5. The late fetal period overlaps that of the infant, with increasing autonomy of physiological function and with further maturity of physical and behavioral characteristics that evoke empathy. During this subperiod the major distinctions between fetus and infant are quantitative and the only sharp discontinuity is birth.

This subdivision of the fetal period deliberately emphasizes characteristics important to the several aspects of individuality and status. These include level of individual independence, of behavior, and of capability for meaningful interaction with others. Each deserves further comment.

VIABILITY AS A FOUNDATION FOR STATUS

Viability, as some capability for individual survival, is an essential attribute of social status. Normal adult members of society assert independence in their behavior and motivation and are valued accordingly. As this behavioral individuality increases during maturation (and declines with aging) social status changes, although a minimal base level is formally maintained from infancy to death.

Therefore, it is reasonable to require some quantum of independent individuality for membership in society. Since offspring in their earliest stages lack both independence and other important aspects of individuality, increasing status in some correspondence with rising individuality is also reasonable. This common-sense approach led the Supreme Court to rely on viability in *Roe v. Wade*.[14]

However, in practice the matter becomes a good deal more complex. For example, the infant has a structurally and functionally independent alimentary tract but remains totally dependent on others for provisioning it. In fact, both the infant and the mother are physiologically and behaviorally "programmed" to shift at normal term from dependent nutrition via the placenta to equally dependent nutrition via the breast. This transition does not signal a change in dependency (actually interdependency), only a change in the means of satisfying it.

But not only does the infant not become independent immediately at birth, it becomes increasingly independent long before birth. And not only has technological life support advanced the *time* of viability, it has altered the *concept* itself. Viability at twenty-six weeks is not the same as viability at normal term. In the range of twenty-four to twenty-eight weeks, viability is *statistical*, with survival increasing and rate of abnormality and subsequent handicap decreasing. Moreover, in this range and even beyond, the whole birth experience is totally changed for everyone involved. It is a technological extravaganza almost entirely under professional custodianship, with family very much in the background. This is necessary under the circumstances, but to say that a premature fetal infant at twenty-six weeks and a newborn at normal term are both "viable" does not make them even remotely equal.

The point is worth emphasis to make clear that physiological viability in the sense of capability for survival is not uncomplicated as a criterion for change of fetal status. The delineation has practical obstetrical utility and affords a convenient medicolegal boundary. But the setting of the boundary is based not as much on the properties of the fetus as on the effectiveness of available technology. Moreover, the criterion is often least useful when it is needed most. For the fetus at the margin of viability, the criterion becomes statistical, as mentioned, rather than individual in its applicability. For a particular baby being delivered prematurely, without assurance of the exact date of conception, its statistical probability of survival is less relevant than its actual level of maturation and the level of available clinical support.

BEHAVIOR AS A FOUNDATION FOR STATUS

Fetal behavior also has its complications as a criterion of status. Behavior is an attractive indicator because it depends on neuromuscular maturation and is directly related to social capabilities. Indeed, it is in social interactions that status is displayed. Nonetheless, the relationship between behavior and status must be approached with caution, as is quickly seen when the significance for status of the earliest human fetal movements are carefully examined. These movements begin between the sixth and eighth week after fertilization, long before the immature brain could possibly invest them with psychic or social individuality. Yet these movements cannot be ignored as irrelevant to status. Movement is a salient feature of human behavior—when we expect it to occur and it does not, that is cause for alarm. When it does occur it is meaningful in indicating the existence of functioning neural circuitry and in anticipating more complex behavior yet to come.

As initiation of overt behavior, movement was suggested in chapter 5 to be a suitable marker for the beginning of the fetal period. The anatomical markers traditionally used are not nearly so indicative of the rising level of maturation of the fetus in comparison with the embryo. Moreover, the initiation of movement is now clinically detectable by ultrasound imaging, and the nature of the movements gives the clinician important evaluative information about the health of the unborn patient. Yet, as mentioned, it would go too far to conclude that movement in the early fetus implies sentience or demands status equivalent to that of an infant. The information now available says that the early movements are based on simple neural circuits such as underlie reflex behaviors—simple, limited in function, elicitable repetitively with little modification, and expressive of neural arcs that do not include the brain. They cannot involve brain mechanisms because at six to eight weeks the brain rudiment has yet to form neurons.

However, the primitive early movements of six to eight weeks after fertilization are elaborated during the fetal period. Ultrasound studies in recent years have provided significant new information. Biologist and ethicist Michael Flower has reviewed and summarized these studies as showing "a transition from simple whole body movements to complex motor repertoires."[15]

Thus hiccuping, isolated arm and leg movements, and head rotation have been reported at eight weeks; stretching, jaw opening, and breathing movements at nine weeks; and swallowing and sucking at eleven weeks. By thirteen weeks, which marks the traditional end of the first trimester, a "basic motor repertoire" has been established. Such movements, more noticeable as the fetus enlarges, are felt by the mother as "quickening" before the midterm of pregnancy, and their continuance gives assurance that the fetus is alive and well. Reportedly an unhealthy fetus will stop moving some twenty-four hours before dying.

BASIC MOTOR REPERTOIRE AND SYNAPTIC FREQUENCY

Flower points out that neurobiologists have found a sharp rise in the number of synapses in the spinal cord as the basic motor repertoire is elaborated. Also detectable are heightened but unorganized electrical activity and increase in the fetus's capability to respond to stimuli, whether these originate within or without. He suggests that increasing fetal activity is likely to result in increasing sensory input from active muscles to the brain stem, and thus to the reticular formation developing within it. Flower sees the netlike neuronal reticular formation in the brain stem as a possible early site of generation of fetal activity.

However, at the time the basic motor repertoire appears in the early fetus, there is no neuronal connection between the brain stem and the rudiment of the cerebral cortex. Moreover, the cortex itself does not yet show differentiated neurons with synaptic interconnection. Therefore, to the degree that the cortex and related higher brain centers are necessary to sentience, no assumption of sentience can be made even at the end of the first trimester, when the basic motor repertoire certainly has been established.

In fact, until twenty-nine to thirty-two weeks the neurological connection between cortex and brain stem does not change substantially. At that time electrical activity of the brain begins to show intermittent patterns resembling some of those seen in normal adults. At this time too neuronal fibers growing outward from the thalamus have entered the cerebral cortex and have made initial synaptic contacts with cortical neurons. Beyond about thirty weeks, the cortical neurons themselves gradually increase their synaptic connections through rich branching and formation of spinelike projections that enhance their available surface for synaptic contacts. At this stage, therefore, the cortex —known to be central to higher coordinative function—takes on for the first time some of the appearance of the adult. Now also electrical activity of the brain shows evidence of maturation, including periodic fluctuations suggestive of cycles of sleep and wakefulness.

It is interesting to compare this obviously incomplete picture of gradual brain maturation with descriptions of the behavior of premature infants, which were provided by Arnold Gesell some forty years ago. He characterized twenty-eight- to thirty-two-week fetal infants as loosely articulated and flaccid mannikins, alternating between brief activity and "limp and torporous desuetude." Torpor, he said, is the "most conspicuous and consistent behavior." The torpor "has little structure or temporal pattern." The fetal infant "is easily roused to brief mild activity, but he is never fully roused. He neither sleeps nor wakes, but only drowses and stirs."[16] This is a graphic account of what might be primitive, transient, and fluctuating sentience.

These neurological and behavioral facts, while still fragmentary and lacking in detail, indicate that the middle of the fetal period is a time of rising uncertainty as to the possible advent of sentience. The available facts speak against the presence of an imaginable state of sentience prior to twenty weeks and for a period of uncertain duration beyond—in all likelihood to at least thirty weeks, when cortical maturation and connectivity noticeably rise. To provide a safe margin against intrusion into possible primitive sentience, the cortical maturation beginning at about thirty weeks is a reasonable landmark until more precise information becomes available.

Therefore, since we should use extreme caution in respecting and protecting possible sentience, a provisional boundary at about twenty-six weeks should provide safety against reasonable concerns. This time is coincident with the present definition of viability, in the context of contemporary life-support technology. The proposed boundary, however, would be based on a substantial neurological rationale relating to intrinsic fetal properties rather than to physiological viability that may shift with technological advances. The designation of twenty-six weeks as a safe barrier against the invasion of sentience conforms to accepted trimester designations. However, this almost certainly will change as more sophisticated and penetrating information accumulates on the time of advent of sentience. That time is far more likely to be later than twenty-six weeks than earlier.

To encapsulate the last three chapters in their application to the status of the unborn, five periods have been recognized as possibly warranting different status. These are the preembryo (zero to two weeks), the embryo (three to about eight weeks), the early fetus (nine to twenty weeks), the middle fetus (twenty-one to thirty weeks), and the late fetus (thirty-one weeks to birth). In the next chapter I shall examine the implications of such division of the unborn period, with emphasis on the major issues that may be affected.

CHAPTER 7

Reaching Decisions on Status for the Unborn

PREVIOUS CHAPTERS have emphasized that, during the development of the unborn, new properties are steadily emerging and individuality, in its several aspects, is rising in a continuing progression. As the developing offspring changes and matures, awareness of its presence and its human quality also increases. Accordingly, the issue of its status—its proper place among us—becomes more and more demanding and problematic.

Should the status of the developing entity change in some correspondence with its changing nature? If so, this might emphasize major developmental transitions as possible markers for successive changes of status. The first objective of this chapter is to examine such an approach more specifically, particularly to highlight the developmental transitions that might merit major status changes. A second objective is to consider a mechanism to implement the approach.

Major Motivating Concerns

It is worthwhile here to restate why the status of the unborn merits reevaluation. The first reason is to articulate the concept of respect for its humanity and to extend that concept explicitly to members of the human species who are in the period of their life history that begins with the fertilized egg and ends with the newborn infant.

The concept of being human, as it has evolved painfully in the adult world (to include all peoples, for example), needs to be appropriately extended as well to the early stages of the human life history. Treatment of the unborn at each stage should reflect the concept that unique values reside in each member of the human family and that protection of those values should be a paramount goal of public policy. The unique human value resident in each individual has been honored and celebrated in the great religions and is reasserted in the secular doctrine of paramount human rights. It and its associated status need to be unequivocally spelled out so that they are applicable to the changing nature of the stages of human developmental history.

The second concern and objective of status definition focuses on the inherent potential of the fertilized egg, and of all subsequent stages of normal development, to become a mature individual in the fullest sense of a person. It is important here to recognize that potential refers to latent but not yet realized properties and characteristics. In this sense, potential is not to be confused with a state of actual being or even with assured realization. For potential to be realized requires further actualizing changes that are dependent on essential enabling circumstances.

Stated another way, potential to *become* a person is not equivalent to *being* a person. Rather, potential is a capability that is contingent for its realization upon particular circumstances (for example, a receptive uterus) that are normally embodied in the human life history. Potential, therefore, is not to be equated with a current existence but with a contingent future. This point is crucial in appropriately assigning status to the unborn.

In these terms, the motivating concern that arises out of the potential of the unborn relates to the welfare of a future person who neither yet exists nor even certainly will exist. How an early stage of development is treated has consequence for the welfare of the person it may become. Concern for the existing early stage is, therefore, only partly for what it actually is now; it is also for what it may become. The latter concern is future-oriented and contingent because what the future may bring is dependent on circumstances that are neither fully predictable nor completely controllable. For example, such anticipatory concern has been rising recently with the increasing risk of exposure of pregnant women to agents that may damage an embryo in the uterus.

This is the subject of the third concern—growing anxiety about deleterious effects during the embryonic period because it is the stage of organogenesis. This complex and not yet well understood process of formation of parts and organs is easily deflected into abnormal courses that do not interrupt life but can lead to later defect or abnormality that limits the quality of life of surviving offspring.

Fourth, as organogenesis advances it brings increasing recognition and identification of the developing entity as human. Recognition of humanness inevitably generates some degree of empathy, a sense of possible sharing of experience that is the earliest active response to an emerging human quality in existence rather than as potential. Such response is precursor to social recognition of a new human being in the full sense—recognition that can lead, for example, to emotional reactions of discomfort, anxiety, and even guilt in situations where pain and suffering may be thought to be imposed on the defenseless embryo or fetus. Whether justified or not, such empathic reactions are the early glimmerings of social interaction and cannot be ignored in considering questions of status.

Fifth, as the fetal period advances, the issue of possible actual inner experience, of sentience or awareness including the possibility of pain and suffering, becomes increasingly pressing. As we lack the ability for direct communication with the fetus, less-certain indirect behavioral indicators gain some credibility. Given their uncertainty, however, it becomes attractive to provide a margin of safety, beginning well before any reasonably imaginable threshold of sentience.

Coping with the Enigma of Human Development

Given these concerns, embodying as they do significant uncertainty and considerable diversity of strongly held opinions, what consensus is possible? Can any status be formulated that may be consensual and yet be able to cope with the special nature of development—its continued progressive generation of complexity with new characteristics and properties steadily arising?

To illustrate this special problem, one may say that developmental progression is less like traveling on a railroad than it is like the course of a wave that forms and then breaks on an ocean beach. A railroad provides a linear sequence of labeled and easily recognized stations; a wave features a rolling rise, a spray-topped peak, and a crashing crescendo as it breaks onto the beach and then quietly dissipates in foam spreading on the sand.

Exactly where are the division points as such waves begin, reach full magnitude, and end? What, overall, are the critical transformations? This is the nature of the problem, which is in fact many times more difficult, in identifying critical transitions in the far more complex process of human development.

To increase the difficulty, a current formulation of status ought not unduly to impede increase of knowledge and further human evolutionary progression (see chapter 8). Can current concerns be met without foreclosing possibly attractive future options?

With all of these difficult objectives in mind, this chapter refocuses on each major phase of unborn development, emphasizing those properties that are relevant to status. Attention is then turned to the central issue, the nature of a decision process about status that might resolve current concerns while remaining open to future new developments. The five phases in the life of the unborn—preembryo, embryo, early fetus, midfetus, and late fetus—flow continuously from one to the other but can be reasonably demarcated by major transitions.

An Individual Cell to a Cellular Individual

A new generation comes into existence at fertilization with the establishment of a *unique genetic individuality* in a *single living and human cell*. This occurs through the fusion of egg and sperm from the two parents. The creative event, following a number of subsequent cleavage divisions, leads to consequential interactions among the product cells, transforming the developing cellular aggregate into a *multicellular entity*. The fundamental biological importance of this transformation cannot be emphasized too strongly; it recapitulates a long and complex evolutionary history that included production of the so-called preembryo as the earliest multicellular mammalian rudiment. That step is fundamental because in achieving multicellularity all additional levels of human complexity become possible.

Accordingly, the status of this early multicellular stage, or preembryo, should be accorded *at least* the minimal respect traditionally paid to cells and tissues derived from the intact human body through surgery or anatomical autopsy. This means essentially that they will not be treated casually as if they were nonhuman material. But the status of the preembryo should go beyond such minimal respect, given its two other characteristics —its particular genetic individuality and its unique potential to become a full person, given appropriate circumstances.

The combination of these two characteristics endows the preembryo with the capability to realize a new human generation in both the biological and the social sense. Its genetic individuality also means that the preembryo has inherent kinship relationships that link it in time not only vertically to parents and future offspring but horizontally as well to give it siblings, cousins, and more distant relatives within the existing human community.

Thus the genetic individuality of the preembryo confers on it a unique value as a new node in a hereditary web that at least theoretically unites the overall human family, both vertically and horizontally. This unity has been reinforced by what we now recognize as shared nucleotide sequences in the DNA of the human genome. The DNA of each of us is astonishingly similar, and we are copycats down almost, but not quite, to the ultimate detail. Our fundamental genetic homogeneity is spiced by only a very much smaller number of dissimilar nucleotide sequences that account for the nonetheless profoundly important hereditary variance and genetic individuality among us.

The point is fundamental. Our new knowledge of molecular genetics supports the reality of the hereditary human web—the term human family has a scientific base. The informational sequences that make up the genetic material of all members of the human species are significantly more than 99 percent identical. Whether in physiognomy, personality, or cerebration, the hereditary differences among us are based on far less than 1 percent of our genetic heritage. The more than 99 percent identity of sequences measures our common kinship.

The preembryo carries that common human message as well as the almost vanishingly small combination of sequences that make it genetically unique. The unique component represents individuality; the vast remainder represents the historical unbroken lineage of the species, evolving from the beginning or beginnings of life. But it must be recognized that genetic individuality leaves the preembryo still far short of the usual notion of a person fully in being, even though it has the important and properly highly valued potential to become one. But if it does become one, it is not just any person—it is a particular one that may be a son, a daughter, a sister, a brother, or perhaps a cousin. Above all, when it has thus been realized it will have a place and a role in family and society, a role formalized in status.

Therefore, though preembryos are not persons, they have value much beyond what is implied by the concept of respect as accorded to the common hereditary humanness of other cells or tissues. The value of a preembryo is highest to genetic parents or to other close kin. But a preembryo can also have very high value to people who are not immediate kin but who wish to bear and raise a child although they are unable to do so through their own reproductive capability.

If the preembryo cannot realize its highest potential as a person—the preferred option—preembryos have other very substantial values. They may provide essential cells and tissues to others. Or they may be the subject of research studies yielding important benefits to all of humanity. In contrast to such possibilities, casual or even ceremonial discard of preembryos amounts to disrespect, in that it fails to appreciate them for the unique entities they are in the hereditary web of the human family.

It is essential, therefore, that choice among the several options for the fate of preembryos not be made lightly nor in the service of narrow or uninformed interests. A consensual framework is required to establish who should make such choices and how. Presumptive rights and responsibility clearly lie with the genetic parents and their kin, assuming that they are willing and able to exercise the rights and assume the responsibility. If, however, the potential of the preembryo cannot be realized within its immediate kinship, some form of broader trusteeship would be desirable. The first responsibility of such trusteeship would be to preserve the continued being and the welfare of the potential person-to-be. But, failing that for any reason, responsibility should lie with the broader community to which the preembryo clearly belongs. The trustee's surrogate role should be exercised as part of this wider responsibility and decision process. Considering the profundity of the issues involved, the process should be at least as carefully considered and designed as that for legal determination of death or the meting out of criminal justice. This matter will be discussed further later.

The Augmented Value in Embryos

Like a preembryo, the embryo is alive, human, and endowed with both kinship and potential. But the embryo has additional characteristics of substantial importance. While a human embryo is developmentally a single individual, normally it is also, somewhat paradoxically, a *functionally dependent temporary part* of its mother. For a period that extends well beyond the embryonic, the embryo and mother are so intimately interrelated that the status of one necessarily impinges on that of the other. The relationship, however, does not involve equal status. The mother is a fully established person with strongly defined rights, while the embryo is rudimentary, both in function and status.

However, if we temporarily set the nature of the embryo-maternal interaction aside and think of the embryo alone, we see that the embryo clearly is biologically quite different from either a preembryo or an infant. As noted, the embryo is alive, human, genetically individual, and with potential to become a future person in the full sense. But it differs from the preembryo in being developmentally single and, especially, in the fact that its structure and function are rapidly changing and emerging into new levels of individuality.

The latter characteristic is the hallmark of the embryo. It is a rudimentary multicellular individual in the full flood of becoming. Its structural parts and organs are coming into existence as it enlarges, and limited function is appearing within them. However, both structure and function remain quite immature, especially with respect to the crucial matter of behavior. What is more central to humanness than behavior? But the embryo remains behaviorally inexpressive and profoundly silent; it is quite devoid of the winning behavior that will later make it so attractive—as a person.

Nonetheless, the rising complexity of the embryo is itself impressive and cannot be ignored when considering its status. The increment in complexity represents gradual transformation of earlier potential into realized and existent value. As realization of potential occurs it steadily heightens the embryo's intrinsic worth and importance.

In part, the added value stemming from realization is contributed by processing and transformation of the inner genetic message. But the added value also represents maternal investment through allocation of nutrients and provision of a protective and facilitative environment. These are hard phrases used by physiologists and economists to describe aspects of growing emotional bonding, what a psychologist might call incipient love.

In any event, the steady augmentation of the embryo's value clearly should be taken into account in considering its status. In fact, the augmented value may register as either positive or negative depending on perspective and circumstance. A wanted child enriches the lives of the genetic parents as development proceeds. An unwanted child is a growing burden. In both instances, however, the increasing realization of potential leads to changing value. This is seen dramatically when a prospective mother who has been ambivalent or negative about her pregnancy undergoes emotional bonding on experiencing the reality of her developing offspring through the technology of ultrasound imaging.

To register in status the overall rising value of embryos, one might consider and refer to them as members of the human family who are in transitional status. They are not yet persons but they clearly are on the way to becoming such; they are semi- or quasi-persons. Thinking of them in this way would validate and dignify the incontrovertible biological statement that human embryos are in a stage of passage, they are in the course of realization, they are fulfilling the life history of the human species. They are, in fact, genesis made manifest.

This transition of the embryo roughly and reciprocally parallels the course of elderly people who are suffering from senile dementia. Some of their accepted characteristics as persons are diminishing or have been obliterated but they continue to be cared about and for, or should be, as fellow-members of the human community. The two situations are not, of course, identical. But each represents a transitional state to, or from, typical personhood. Elderly people with severe cognitive defect might not be admitted to personhood if their status were to be reconsidered on strictly definitional grounds. But by general consensus and for complicated reasons, their status is continued as an entitlement.

Similarly, it would be premature to grant rights as a person to an entity so incomplete as an embryo, but one can reasonably argue that it is much closer to being one than a preembryo or an unfertilized egg. To formally entitle the embryo to status as a member of the human community, but not yet a person, would signal recognition of its rising value and fundamental continuing potential.

What would be gained concretely and practically by such status? Perhaps most significantly, a greater awareness among us of the emerging entity and its nature. For most people, embryos are quite unfamiliar, not only in presence but in their very character. For example, adolescents do not recognize growing sexual interests and capacities as enhanced ability to produce an embryo. Is it naive to suppose that if the embryo and its status were more definite in their minds they would be more motivated to avoid unwanted pregnancies and contraceptive abortions? And might such effects extend even to postadolescent sexual responsibility as well?

Such status might also encourage development of abortion procedures that are less lethal for the embryo. Current procedures are designed for the safest possible termination of the pregnancy for the mother, an entirely reasonable approach if the embryo is assumed to have no value. But if pregnancy could be terminated equally safely and satisfactorily without destroying the embryo—and if embryo survival were seen as desirable—the embryo would become a more valued member of the human community. Certainly such a change would show far greater respect for embryos and their human relationships than does current obliteration as if they have no value whatsoever.

To move in this direction, however, certainly raises complex moral and public issues, undoubtedly requiring long deliberation, debate, and accommodation. It would also require an institutionalized system of oversight to assure that all activities involving extrauterine human embryos were carried out within sanctioned objectives, by authorized and competent personnel and in accord with promulgated ethical standards. Oversight of this kind might be provided under several existing precedents, about which more will be said later.

For the moment, the central point is that the embryo, though increasingly to be valued, remains too immature structurally, functionally, and behaviorally to warrant classification as a person. It should, however, be recognized as a member of the human family entitled to protection against treatment or deprivation of life that is casual and demeaning.

The Fetus and Its Mounting Individuality

At about eight weeks, the embryo merges into the fetus, a phase of rapid growth and maturation that is especially characterized by the onset of movement as overt behavior. Movement is critical because it signals behavior, in turn significant to social status. Also, movement is dependent on, and therefore diagnostic of, the level of neuronal maturation in the central nervous system.

The fetal stage is one of generally increasing individuality, both functional and behavioral, and leads to capability for external survival (viability) by the end of the second trimester (twenty-six weeks). Bridging the long duration between the embryo and the infant, the fetus is by far the most difficult period to deal with in terms of status. The difficulty cannot, in fact, be resolved by simple categorical definitions. Rather it must be viewed for what it is—a prolonged transitional and gradual maturation requiring flexibility both in definition and in its assessment in particular cases.

How should the status of the fetus differ from that of the embryo that precedes it and the infant that succeeds it? In comparison with the embryo, the fetus is enlarging substantially, is increasingly active behaviorally, is gaining autonomy of function, and is increasingly identifiable as human from its external characteristics. In turn, these changes engender rising feelings of empathy and compassion. When confronted by a fetus, people are increasingly likely, as the period advances, to express social recognition in their emotional and behavioral responses.

In comparing a fetus with an infant, level of social interaction is again a critical difference. The early fetus, although behaviorally active, is only minimally responsive to its environment and not at all attentive to it. Even the prematurely born relatively late fetus is only sluggishly responsive. The infant born at term, on the other hand, clearly is on the verge of social interaction. Its capability in this direction is anxiously anticipated and solicited by its caretakers as an indicator of its well-being. Responsiveness, even when only intermittent, is a source of delight to them.

Interactive awareness, therefore, is rapidly rising as normal birth approaches. Neonatologists' recognition of this is manifested in the difference in arrangements and decor between an intensive care nursery (ICN) for the prematurely born and a normal neonatal nursery. The ICN is as stripped down in amenities as the initials that designate it. Amenities and social esthetics are Spartan and clearly secondary. Naked fetuses lie flattened in decidedly unflattering submission to the gadgets surrounding and pressing upon them. It is clear that the fetal infant, if not oblivious to its social environment, is not expected to be seeking much satisfaction in ornamental interests.

Status and Inner Awareness

Existing biomedical knowledge does not require an assumption of significant inner awareness of an environment during the course of fetal life. In the early fetus there certainly is not an adequate neural basis for inner awareness, as judged from the developmental neurology of the brain. Although the same cannot be said, with equal conviction, of the late fetus, there is no compelling evidence that the statement is not also applicable.

Nonetheless, until more decisive evidence is available, prudence calls for error on the side of caution. Thus, to avoid violating possible inner awareness, fetal status might be altered at the end of the traditional second trimester, roughly at twenty-six weeks. This is close to the time of statistically reliable viability with existing life-support technology. It is some four weeks later than the earliest demonstrated thalamocortical connections in the developing brain but some four weeks prior to the first maturational change in the brain's electrical patterns and to extensive linkages among cortical neurons. The exact point should be subject to adjustment as more precise data become available. The major impact would be that presumed inner awareness rather than viability would then be the key criterion for the definition of status.

Few arguments have been advanced against a strongly protective status for the unborn once inner awareness is definitively present. Rather the problem has been uncertainty about the time this occurs, as judged from objective data. A similar uncertainty exists with respect to inner awareness in animals. A reasonable approach requires encouragement of research leading to more adequate knowledge, and such approaches can now be seen on the horizon. Meanwhile, adoption of a conservative posture in providing a margin of safety for protective status seems prudent.

Under these assumptions, for the human species, the status of the fetus (whether *in utero* or *ex utero*) should not exclude the possibility of inner awareness beyond twenty-six weeks of gestation or its equivalent level of maturation. Therefore, the third-trimester late fetus should be protected against traumatic intrusion into a possible state of inner awareness.

Such a protective status would not be appropriately assigned on the basis of fundamental rights (for example, constitutionally) but as a judgmental entitlement based on current knowledge. The correspondence of the selected time to that of viability, given existing technology, should be regarded as currently convenient but not necessarily permanent. Thus, even if technology should advance further in allowing early fetuses, or even embryos, to be externally viable, this need not require them to have the same status as late fetuses.

Achieving and Implementing Unborn Status

To this point, I have emphasized the complexity of the problems of status for the unborn and the general directions that might be taken to cope with the problems. I now turn more specifically not only to how the problems might be resolved but to how such resolutions might be effectively implemented.

Status for the unborn clearly is a matter of public policy and at the level of broad significance and importance. Although there may be little agreement on anything else, each of the polar views in the abortion controversy has sought to establish its position at the highest legal level, whether by interpretation of the Constitution by the Supreme Court or by amendment of the Constitution itself. Clearly, those who regard the concept of humanity as fundamentally important believe that questions about human generation require the highest level of deliberation and decision.

A first step toward gaining consensus on so fundamental a matter is to explore fully both its substantive core and its ramifications. This means that the issues both directly and indirectly involved, the positions held by various parties, the rationales of the parties, and the accommodative options that have or might be considered have to be thoroughly canvassed. Such a preliminary analysis necessarily also entails review and wide sharing of relevant facts, as supplied by the best-informed and most reliable sources. The process should be carried out in a national forum of appropriate visibility and stature, to which all interested parties have access and by which all concerned can be informed.

Reaching Decisions on Status for the Unborn

The nature of the forum is a matter of major significance and interest to all concerned. Of the various devices that have been utilized or imagined in related areas, one of the more successful in the United States in recent years has been a national commission. Several of the most effective such commissions have dealt with issues similar in nature to the status of the unborn and have touched on the issue itself. These include the National Commission for the Protection of Human Subjects of Biomedical and Behavioral Research[1] and the President's Commission for the Study of Ethical Problems in Medicine and Biomedical and Behavioral Research.[2]

These successful commissions were initiated and funded by the Congress for extended deliberation over three years. Their membership was broad in scope, including professional and laypeople of differing perspective; they were supported by professional staffs; they held public hearings; and they published multiple reports that set out general considerations and rationales as well as specific recommendations. Their impact on subsequent policy formation was consequential and salutary. Such a commission could be helpful in focusing the process of definition of status for the unborn.

It is not, however, the only possibility. In late 1985 the Congress established within the legislative branch a Biomedical Ethics Board to "study and report to the Congress on a continuing basis on the ethical issues arising from the delivery of health care and biomedical and behavioral research."[3] The board is to have a professional Advisory Committee to guide it in its studies and other activities. The board's structure is parallel to that of the Office of Technology Assessment, which for over a decade has advised the Congress on scientific and technical matters requiring national policy and possible legislative attention. The Biomedical Ethics Board could be an appropriate focus for national deliberation on the status of the unborn. Other possibilities should be explored as well.

The essential requirement for any such mechanism is that it

be of credible national stature and that it be dedicated to the broad public interest. It is also desirable that the body be exclusively deliberative rather than administrative, that it operate in full public view, that it be given a specifically formulated agenda for decision and implementation, and that its conclusions be widely circulated and debated before being implemented.

The full content and priorities of the agenda, the recommendations to be made, and the details of implementation will, of course, all emerge from the actual deliberative process. But certain patterns may be recalled from recent experience in related areas, patterns that appear to reflect new trends in public decision making in heavily value-laden technological controversies. A useful case in point is the recombinant DNA controversy of the middle and late 1970s.

The Model of Recombinant DNA

In so complex an area of public policy making, any "model" can be questioned as to its pertinence in detail. Nonetheless, the recombinant DNA controversy seems to have set certain useful precedents for the matter of status of the unborn. This is because a seemingly sudden and certainly dramatic advance in molecular genetics not only posed substantial policy questions of immediate safety but raised profound issues of longer-term impacts on the earthly environment and control of human evolution. Some posed the issues in such apocalyptic language as "contamination and pollution of the human genome" and "playing God."[4]

These concerns were projected into the general community at a time of considerable uncertainty about a number of the technical questions involved. As initially an accepted and appropriate national forum for deliberation and debate was lacking, diffuse and largely unstructured clashes occurred among people imagining sharply different speculative scenarios. The atmosphere became confrontational between those who were for and those who were against the general concept of "genetic meddling." For several years the controversy raged in local communities, state legislatures, and the halls of Congress.

The emergence of national focal points for agenda setting and debate seems to have contributed strongly to a more constructive course. The process began with an improvisation—the establishment in 1974 of a Recombinant DNA Advisory Committee (RAC) to the director of the National Institutes of Health (NIH). This was reasonable because the NIH was the chief supporter of the molecular genetic research that gave rise to the problem. The RAC became the center for public discussion of appropriate risk-mitigating guidelines for continued research. As information and opinion were mobilized and experience accumulated, informed estimates of the range and level of risk declined. By recommendation of the RAC, the specific guidelines, by then largely administered by local committees, were gradually attenuated through public RAC debate and recommendation. Today most molecular genetic research is free of limitation, and there has been no known case of biological or medical harm through use of recombinant DNA technology.

On the other hand, as industrial interest in the new technology has grown without full resolution of risk questions at industrial scales of operation, RAC temporarily assumed broader responsibility, for example, for deliberate release of recombinant products into the general environment. More recently, agencies with defined regulatory responsibility, which NIH does not have, have been parceling out the regulatory task, leaving RAC to function only in the area of biomedical research.

The Emergent Policy Pattern

The pattern that somewhat painfully, but now seemingly successfully, evolved includes: (1) exercise of national governmental oversight through a quasi-governmental advisory body (RAC); (2) delegation of the general national authority to local supervisory groups appointed by officials of local institutions that are directly involved; (3) responsibility placed on immediately involved professionals carrying out activities under general guidelines; (4) sorting out and assigning appropriate regulatory responsibilities to existing regulatory agencies, such as the Food and Drug Administration, the Environmental Protection Agency, the U.S. Department of Agriculture.

Similar multitiered mechanisms operate in assuring appropriate standards in the use of animal and human subjects in research. At least theoretically, the structure allows federally established policy oversight to be exercised without the "heavy hand" of direct bureaucratic intervention. At the same time it affords local, "on-scene" decision making with opportunity for public access and discussion through publication of key proceedings in the *Federal Register*.

The pattern contains useful precedents for defining and implementing status for the unborn, with detailed operational decisions made by local evaluative groups within the context of an overall policy. In such a tiered concept, a broadly constituted national group—advisory and responsive to governmental authority and responsibility—would formulate guidelines and maintain federal oversight. The oversight group, possibly a Commission on the Unborn (CUB), would supply general definitions of status, guidelines for specific decision making, and continuing oversight and evaluation of the resulting activity.

Local groups would operate within CUB definitions and guidelines to participate, as necessary, in making judgments about individual cases in specific circumstances. Local groups might be existing committees on the use of human subjects in research or hospital staff committees, suitably redefined, or they might be committees specifically constituted to deal entirely with decisions about the unborn. The arrangement would provide national standards while allowing a degree of flexibility for local option and innovation.

For example, a local group might receive a request to transfer a frozen preembryo to the uterus of a non–genetically related mother. The national guidelines, promulgated by a responsible federal official on advice of the CUB, might state that such a transfer requires reasonable assurance that the resulting infant will be reared in a stable family unit. Candidates to receive the preembryo might include a married woman who wishes to rear the resulting child with her husband (which would be a case of prenatal adoption) and an unmarried woman who has made an arrangement to deliver the infant to its genetic parents (with the woman serving as a surrogate mother). The local committee would be free in its recommendation to give preference to one or the other, depending on local standards and priorities.

Or a research proposal might be submitted by a group of qualified investigators who require twenty preembryos to fulfill a study plan. The national guidelines might specify that research utilizing preembryos can be carried out only to increase the success rate of IVF and then only if the matter cannot be tested in any other way. A given local institutional committee might or might not agree to assign one or more preembryos for the research, depending on the priority the committee assigns to investigation (even if otherwise appropriate) in comparison with clinical demand for embryo transfers.

As a final example, such a local group might serve as an appeal body in difficult cases involving termination of unusual life support for prematurely born infants afflicted by substantial structural or functional defect.

The research scenario just described raises yet another issue.

If the unborn are appropriate subjects for research—a matter to be determined at the national level—who can carry it out? The 1985 report of the British Warnock Committee extensively discussed and commented on the matter with respect to preembryos. It recommended that such research be permitted only by groups specifically licensed to provide IVF.[5] Specific licensure for such purposes has not been usual practice in the United States but, given the sensitivity surrounding human embryo research, some policy assuring fully responsible conduct seems essential.

A general issue raised in England is whether any authority other than governmental can provide a basis for status of the unborn. Despite the formal parliamentary inquiry that resulted in the Warnock report (which recommended licensure for all facilities and persons involved in IVF), legislative maneuvering in the tense atmosphere created by the abortion controversy led to postponement of governmental action on the Warnock recommendations. Rather, a voluntary professional group, the Royal College of Obstetrics and Gynecology, has set the standard.[6] A somewhat similar effort is being made in the United States by the American Fertility Society.[7]

While voluntary groups can undoubtedly provide some measure of constructive medical standard-setting, the process obviously cannot achieve the objectives of full public decision making. Moreover, so far the procedure has not been tested as an approach to so complex and sensitive a matter as defining status for early human life.

Making Decisions About Life

The difficulties of making decisions when newborn human life is at stake in intensive care nurseries have been sharply depicted by Anthony Rostain.[8] As noted in chapter 6, this often involves judgments about whether to initiate or continue life-sustaining treatment for fetal infants when it is clear that they are doomed to early death or fundamentally unsatisfactory lives. Rostain describes the dynamics of decision as "complex and often subtle." These dynamics occur in a diverse group of people that may include an attending neonatologist, medical fellows and residents in training, nurses, social workers, parents, and sometimes other consultants.

The discussion may begin at morning rounds with a report from a resident about an infant's difficulties the preceding night. The resident may raise question about the infant's "code status"—that is, what should be done in the event of a cardiac or respiratory crisis requiring emergency treatment. "The subsequent flow of conversation will depend a great deal on the tacit understandings shared by the team." If the outlook is dismal, "the group will focus on specific tactics for approaching the parents or for deciding on the means by which treatment will be withheld."[9]

Rostain's revealing low-key account of the decision process describes how differences within the group are discussed and usually brought to consensus. "Once a consensus is reached concerning the decision, and a means of forgoing treatment has been agreed on, preparations are made for the infant's death." A "sad and solemn ritual" follows. Participants share the sentiment that, despite the sadness, "it is better this way."[10] It is noted, however, that until "institutional changes are enacted . . . current participants will continue to bear the burdens of making decisions in a climate marked by increasing uncertainty, fear, and mistrust on all sides."[11]

A Designed Status

Is there, or should there be, a better way? Should groups like the one described bear so great a burden of decision without more clearly defined and promulgated public guidelines? What might guidelines defining status for the newborn say that would be helpful? One cannot forecast the result of what should be a public deliberative process; but the following is an example of what might emerge from such a process, based on the preceding chapters of this book.

Human society is taken to consist of individuals with rights and responsibilities that are assigned through the mechanism of societal status. Each social individual, however, goes through a biological life history involving fundamental developmental changes from a single cell to a complex multicellular adult. There is no *a priori* reason why rights, responsibilities, and status must remain unchanged throughout this evolutionarily established and complex biological life history. It is only necessary that these social characteristics be carefully and appropriately specified in relation to the biological life history in ways that all can understand and respect.

Birth has long been accepted as the fundamental point of initiation of a societally defined individual. But status does not remain constant from birth onward—it continues to change in some measure until death. Similarly, there is no reason why changes of status should not occur during prenatal life as our knowledge of this period grows.

The earliest change that is relevant to status clearly occurs at fertilization, when genetic individuality, including kinship, is established. A second major change occurs with onset of formation of the embryo as the rudimentary new individual. A third is the transition of the embryo to a fetus, marked by active movement as the precursor of interactive social communication. During the fetal phase there is increasing possibility of a fourth major change, the onset of sentience or inner awareness.

By best available criteria, the early fetus (up to twenty weeks) is certainly presentient, as is almost certainly the middle fetus from twenty to thirty weeks. But current knowledge cannot fully exclude minimal sentience during the late fetal phase (thirty weeks and beyond), though most evidence is clearly against it in at least the first several weeks of the phase. Moreover, the late fetus may be born prematurely and survive as an infant if given intensive technological care.

During this entire unborn period, the developing offspring is biologically human and therefore deserving of at least the first level of human status—respect for and protection against denigration of its human character. But from fertilization on, it also commands a second level of status. It becomes unique genetically and can claim kinship within the human hereditary web. Moreover, it possesses potential to become a particular person in the full sense. The combination of these characteristics provides the basis for definition of the status of the preembryo.

A third level of status may be justified for the embryo. Besides containing preembryonic characteristics, it is now a single individual in actual active formation of parts and organs. Part of the earlier potential of the preembryo is now being steadily realized as organogenesis proceeds and the embryo becomes increasingly complex. Accordingly, it should be increasingly valued and protected as a member of the human family, though it is still considerably short of a person in the full sense.

With the onset of primitive movement as a first component of behavior, the fetus comes into being. Behavior is the foundation of social interaction and, therefore, its very occurrence might itself warrant change of status. But, though behavior evokes empathy, it need not and should not be equated with inner experience. Nonetheless, the onset of behavior and the neural maturation it reflects make assignment of status in the fetal period especially problematic.

There is no sound scientific basis for assuming that inner experience exists in the fetus prior to the third trimester of pregnancy (twenty-six weeks). A highly conservative policy might, in these terms, give to third-trimester offspring, either unborn or prematurely born, the same status as is extended to infants born normally at term. However, the status of the entire perinatal period should be reviewed regularly in the light of advancing knowledge. Advancing technologies that are expanding options and choices at the threshold of independent existence may entail frequent need for new means of decision making and new definitions of the roles of the parties involved in perinatal decisions.

What seems clear is that treatment of the unborn at all stages, including their use in investigation and research, should be under general guidelines provided by institutionalized societal oversight. The guidelines should ensure vigilant protection of humanity, in its broadest sense, against denigration or exploitation from any source. This is so fundamental a human concern that it cannot be left to either casual improvisation or heavy-handed doctrinaire authority. It calls for the application of consistently insightful human wisdom, generated in some form of continuing broad conclave.

CHAPTER 8

Epilogue

I CANNOT LEAVE the subject of the unborn without expressing some thoughts in an even broader perspective. The close approach not only of a new century but of a new millennium invites apocalyptic visions. To escape the nearsightedness of the contemporary, one must imagine futures—not to prognosticate but to see the present in better perspective. What will the issues and events of today look like at the end of the twenty-first century or, even more difficult to answer, at the end of the new millennium?

Given such passage of time, how will the placement of a relatively small number of preembryos in frozen storage be viewed? What would be the impact *then* of transferring *now*-foreign genes into a necessarily small number of human zygotes among the huge number annually produced? With our total human population climbing rapidly above 5 *billion*, what difference will a few thousand, or even a hundred thousand, frozen or genetically altered preembryos make in the long term?

In fact, from our vantage point, such procedures as frozen storage and gene transfer, unless highly concerted and focused, are very unlikely to affect the future of our teeming human billions. But there is a scenario, and one that is not entirely fanciful, that has a more consequential outcome. This epilogue deals with that scenario as a mind-expander.

The first human footstep was set on the moon on July 20, 1969, and the first landing of an Earth-made device on Mars occurred on July 20, 1976. I remember these dates well because both seemed to me to be welcome presents on my birthday. Today there are hundreds of satellites in orbit around the earth, and one such device is in the far reaches of the solar system— about to move into extrasolar space as a "message in a bottle" to whomever may be beyond.

Moreover, here on Earth there is serious discussion of targets for space exploration and at least temporary habitation "out there." One proposal is for a landing party on Mars, or one of its moons, early in the coming century—possibly mounted jointly by the United States and the Soviet Union. And a conference at the Los Alamos National Laboratory, nidus for the atomic bomb almost a half century ago, seriously discussed prospects not only for extraterrestrial colonization but for extragalactic migration![1]

What does such speculation have to do with the status of the unborn? A little arithmetic of extraterrestrial colonization will help. Travel to the moon, using current technology, took a small number of days each way—a long distance to commute but no more disruptive to family life than a tour of duty on a navy carrier or submarine. However, commuting to our neighbor planet, Mars, is another matter. It would take some nine months each way, under travel conditions considerably short of a luxury cruise. That is not an easy way to start or raise a family.

And Mars is close among extraterrestrial targets. It took only nineteen minutes for radio signals from the Mars lander to reach Earth. Similar signals from the outer planets would take hours or days, with human travel times increased in proportion. But distances and times do not become really impressive until one wanders out into the "local" galaxy. The nearest neighboring star is some four light-years away; flight times for current human-carrying vehicles would be of the order of an average human life span. And our "home galaxy" is about 800,000 light-years across its disc! That is a fair distance even for a Methuselah traveling at the speed of light.

This kind of arithmetic becomes staggering when one turns to intergalactic travel, yet it has not deterred intrepid exploratory imagination. Given all the time in the world—or rather in the cosmos—it probably will not preclude less ambitious extraterrestrial colonization. However, it will require a new look at human reproduction, which did not evolve to meet the exigencies of space travel and colonization.

Space colonization may well have consequences for the status of the unborn, because of the logistics and economics of space travel and the biology of small founding populations. The first places a premium on small numbers of travelers and the second on wide genetic variation. A colony of travelers even on nearby Mars would be substantially isolated reproductively from the reproductive population on Earth.

Small sexually reproductive populations of any kind have greater difficulty maintaining themselves over generations than larger ones because inbreeding is likely to bring out unfavorable recessive characteristics. On Earth this is a recognized problem in the biology of small islands and of attempts to rescue endangered species such as the California condor. Human founder populations, whether on space arks or stable extraterrestrial sites, will have to carry more genetic variation than will be allowed by the logistics of transit of persons. Stated biologically, they must carry more genotypes than phenotypes. Enter the unborn in the frozen state.

Transport of frozen preembryos would minimize required space and resources and, if the preembryos were carefully selected from Earth's genetically diverse human population, could enlarge and diversify the gene pool of the new colony. Such a conception, in both senses of the word, is not without difficulties and complications. Among other things, it presumes that preembryos might be created to be transported and to be raised by foster parents in a distinctly strange land and with very little in the way of informed consent. But, to be sure, Adam and Eve are said to have had little more in the way of security of descent.

I repeat that such a scenario is not a prognostication. It is intended to emphasize that the future may provide a different context for evaluating reproductive options. Whether frozen embryos may be on the shipping manifest for early colonization of Mars is dubious. But that planning of such missions will have to include consideration of what is necessary for reproductive success is *not* dubious. Just as space migration takes more than a covered wagon, successful planting of a colony will take more than just bringing along the kids. Reproductive technology will have its place, and it can be expected that the more options it provides the greater will be the likelihood of success.

Thus what research is now being done to meet the needs of subfertile couples on Earth (and questioned on that account as to what its priority should be, given the costs) may later be found necessary to avoid the loss of subfertile extraterrestrial human colonies. Perhaps the underlying research should be supported not only by the National Institutes of Health (which it is not, largely because of the tortured politics of abortion) but by the National Aeronautics and Space Administration.

One can go further along these lines. By the time the first colonization is attempted, even only as far out as Mars and its moons, the human genome is likely to be extensively mapped and revealed in much detail if not in its totality. Many genes for critical hereditary characteristics, both normal and defective, will literally be in hand. Intensive research is now in progress attempting to cure genetic-defect diseases by transferring a normal gene to function for a defective one. The motivation is to save the blighted lives of tragic infants, born to live but not to thrive. The effort meets the highest standards of right-to-life aspiration.

Epilogue

If success is achieved, similar technology is likely to be applicable to alleviating subfertility, not only on Earth but in colonizing Martian moons. Research is already pointing to the possibility that some couples are subfertile not because of either partner's individual capacity but because of a hereditary incompatibility between them that is somehow like cell and tissue incompatibility in transplantation. If further research is successful in demonstrating and overcoming such incompatibility, the resulting technology might reduce the survival problems of small founder populations.

But the space colonization scenario raises a still broader question. Some three thousand human genetic diseases have now been catalogued; individually most are quite rare, but together they affect a significant number of people. It is estimated that each human being carries perhaps three to five genetic defects in a recessive state. In selecting a founder population for extraterrestrial colonization, a number of factors will be considered. Should one factor be the result of screening applicants for the number and nature of genetic defects? Would it be appropriate to eliminate otherwise qualified candidates in order to reduce the genetic burden in the founder population? Would this be an objectionable practice of eugenics?

Before answering too quickly, another possible scenario of the future should be mentioned. Most people earnestly hope that a nuclear war will never occur and that accidents such as the recent one at Chernobyl will not recur. Yet neither possibility has been firmly excluded, and each may be expected to leave a "genetic scar" in exposed populations, a legacy of altered genes that usually are detrimental. The only ways currently known to avoid propagating such scars to subsequent generations is to control the reproduction of affected individuals or to transfer normal genes to them. One asks again: Would this be objectionable eugenics?

However these specific issues may be resolved, it is clear that the most precious cargo carried by a colonizing space vehicle will be its human genetic heritage—the product of millennia of interaction between life and Earth, now to be tested in unearthly environments. The heritage may be embodied in carefully selected reproductive adults, or in frozen preembryos, or perhaps even in "spare parts," lengths of DNA for later transfer of particularly important stored genes. These would, indeed, be the "colonial jewels." In such speculative future scenarios, biological engineering joins its more traditional and better-developed physical and chemical confreres in providing foundation for the human prospect. Pertinent to the success of such enterprises will be the cultural context from which they may be launched. Habits of thought and systems of belief carried forward unchanged from a frozen past will limit prospects that may extend beyond any now foreseen. The special difficulties in dealing with this problem in connection with human reproduction were clearly apparent in previous chapters.

It should also be noted that, though the casual eye may see it otherwise, it is not happenstance that space technology and biotechnology surged forward in the mid-twentieth century. The surge derives from the close and mutually supportive interaction that has evolved between advancing science and technology—especially manifested in the rich connections and cross-feeding among the diverse technologies derivative from science and now dominating our culture. The integrative nature of this technological web provides the link between the almost simultaneous, and seemingly fortuitous, burgeoning of otherwise largely unrelated space and biological science.

Thus the linkage is not fortuitous, but it is fortunate. The earthly stew, within its existing confines, is showing signs of unhealthy fermentation and rising pressure. The pressure might be relieved, before it becomes explosive, by broadened perspectives that direct it outward into the openness of space. Blind aggressiveness might thus be sublimated before vicious reciprocations turn into total destruction.

Epilogue

Let us return to the now relatively small case with which this book began. The unresolved contention over the status of the unborn is frustrating the advance of knowledge and its application to human reproduction. In turn, in the larger context, this frustration limits other interacting areas, including perhaps the provision of effective means for space colonization. The scenarios of this chapter provide a homily which says that our problems and opportunities are, to some degree, all of one piece. Ignorance can be charming, intelligence can be sold in the marketplace, and reciprocal terror can be portrayed as pursuit of stability. But catastrophe and wisdom are each, and together, of one piece.

It is in such a context that we must strive to see the unborn more clearly. The enigma of their status encapsulates the struggle to recognize and value individuality and yet achieve and maintain the reality of collective humanity. Like the unborn, the collective human phenomenon may be still greater in its potential than so far revealed. In this enlarged frame, humanity spreading from the earth is a new birth. If so, all we have so far been may be as if we are yet unborn. Perhaps we have yet to discover what it means to be fully human. Would that be playing God? Perhaps it would be only seeking to learn how.

APPENDIX

NOTES

INDEX

APPENDIX

THIS APPENDIX contains figures referred to in the text as well as additional background information on human development that may be helpful to readers. The figures, with modifications, are from textbooks on human embryology and neurology.

FIGURE 1
Fertilization and Cleavage: First and Second Week

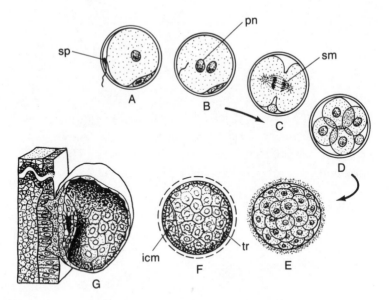

In *A*, a single sperm (*sp*) has penetrated the envelope of the egg and lies in contact with its surface. The egg nucleus is visible near the center of the cell. In *B*, the sperm has entered the egg. Its discarded tail is visible near the point of entry and its head has transformed into the paternal pronucleus (*pn*) approaching the maternal one. In *C*, the zygote is undergoing its first cleavage division (mitosis) with intermixed maternal and paternal chromosomes moving toward the poles of the division "spindle" (*sm*) and the bulk of the cell pinching into two halves. In *D*, a second cleavage division has occurred, yielding the four-celled stage. In *E*, additional cleavage devisions have produced a packet of cells of substantially reduced size—the morula (little mulberry). In *F*, a fluid-filled cavity has appeared in the now-enlarging packet of cells, and two cell populations can be distinguished: peripheral cells, which will function as trophoblast (*tr*), and the small inner cell mass (*icm*), which is precursor to the embryo. Meanwhile, the preembryo has shed its outer noncellular envelope. This stage is the blastocyst, which by now has arrived in the uterine cavity after transit through the oviduct. In *G*, the blastocyst is in contact with the uterine wall (early implantation), the trophoblast has started to penetrate the uterine lining, and the inner cell mass has begun to form a cavity of its own. At this stage there is not yet a single embryo; twinning may still occur.

NOTE: Adapted from R. Dryden, *Before Birth* (London: Heinemann Educational Books, 1978), p. 9; and K. L. Moore, *The Developing Human: Clinically Oriented Embryology*, 3rd ed. (Philadelphia: W. B. Saunders, 1982), p. 34.

Appendix

FIGURE 2
Blastocyst at Implantation

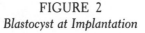

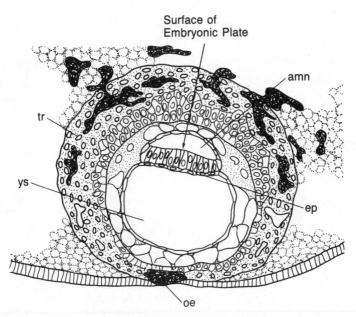

An implanted blastocyst, at ten to twelve days after fertilization, as seen in a thin slice cut through the original entry point (*oe*). The trophoblast (*tr*) is now quite extensive and has irregular blood-filled spaces (black) linked to those in the maternal tissue. Meanwhile the inner cell mass has formed two cavities that are precursors to embryonic structures—the amniotic cavity (*amn*) and the yolk sac (*ys*). Between the two is the embryonic plate (*ep*), the precursor to the embryo itself. It is a flat plate of columnar cells with an associated thinner cuboidal layer. Soon the plate will become three-layered through activities centering in the primitive streak.

NOTE: Adapted from R. Dryden, *Before Birth* (London: Heinemann Educational Books, 1978), p. 10.

FIGURE 3

The Preembryo in Relation to the Mother

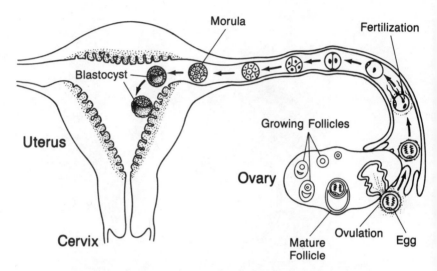

The diagram shows the path followed by the egg and developing preembryo from the ovary (lower right) into the open end of the oviduct or fallopian tube, where the egg may be fertilized by sperm. The zygote continues along the oviduct (while cleaving) into the uterine cavity for implantation.

The egg (usually one per menstrual cycle) has matured in a growing ovarian follicle made up of "nurse cells." A mature follicle ovulates at each midcycle. Nonfertilized eggs are discarded in the subsequent menstrual flow, if not sooner.

If the egg is fertilized, complex hormonal changes in the brain stem, pituitary, and ovary (initiated by a signal from the preembryo) prepare the uterus for implantation and prohibit sloughing of its lining in menstruation.

NOTE: Adapted from K. L. Moore, *The Developing Human: Clinically Oriented Embryology*, 3rd ed. (Philadelphia: W. B. Saunders, 1982), p. 36.

Appendix

FIGURE 4
Primary Organization: Embryonic Axis and Neurulation

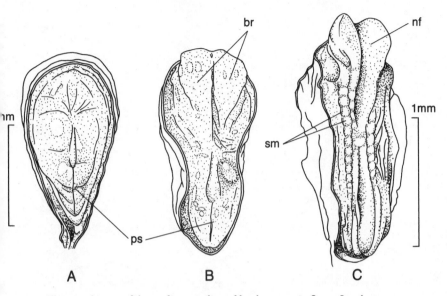

The view here is of the surface—indicated by the arrow in figure 2—shown after removal of the surrounding membranes. Note the one-millimeter scale as an indicator of size as the axis appears and extends.

A. Surface of the embryonic plate at about two weeks after fertilization. The linear depression at the narrower end of the plate is the primitive streak (*ps*), the site of cell migration inward to form a third embryonic cellular layer that spreads sideward and ahead of the primitive streak. The stage shown is when the head-to-tail axis of the embryo-to-be is being established, marking the beginning of the embryonic period.

B. Midway in the third week, the embryonic axis is now apparent, marked by the two bulges that are the first manifestation of a forming brain (*br*).

C. At the end of the third week, the embryonic axis is clear with the still-open neural folds (early brain rudiment) (*nf*) at the upper head end, fusion of the neural folds beginning at the middle level, and segmented cellular masses (somites) (*sm*) forming from head to tail on either side of the neural plate as the axis extends.

NOTE: Adapted from R. Dryden, *Before Birth* (London: Heinemann Educational Books, 1978), pp. 11, 12.

FIGURE 5
Embryonic Developmental Course: External Features

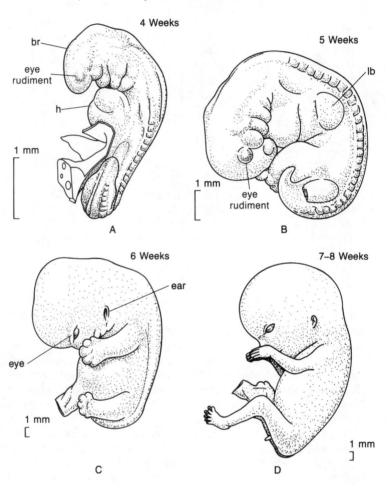

A — 4 Weeks
br
eye rudiment
h
1 mm

B — 5 Weeks
lb
eye rudiment
1 mm

C — 6 Weeks
ear
eye
1 mm

D — 7–8 Weeks
1 mm

Appendix

A. Late in the fourth week. Note the rapid growth, indicated by the one-millimeter markers. The head, tail, and major musculo-skeletal axis of the embryo are now clearly distinguishable. The head end bulges due to the forming brain (*br*), and the beginning of eye formation is indicated by a secondary oval bulge from the brain (*eye*). The large prominence in the midsection of the embryo is due to the heart (*h*) (and liver), which initiates its lifelong beat at about this time. The cutaway in the lower left has severed the connection to the trophoblastic exchange surface with the uterus (placenta).

B. Early in the fifth week. Note again the rapid growth as indicated by the one-millimeter marker. The brain area continues to expand, the eye rudiment is clearer, and limb buds (*lb*) are prominent on the flanks, although they are still without recognizable detail. Note the tail rudiment, which will disappear in later stages.

C. Mid-sixth week. The eye is now distinct and a slot marks the beginning of the formation of an external ear. Note the relative size of the head area, blocked out but not yet clearly separated from the rest of the body. Note also that the limb buds have elongated and now show five protuberances on their paddlelike ends.

D. Seventh to eighth week. The head is increasingly delimited by the forming neck, and hands and feet are distinguishable from each other and have humanlike characteristics. Thus, as the embryonic period closes, the early fetus is recognizable as to species. However, it is less than an inch long and is highly immature in overall body conformation and in both structural and functional detail.

NOTE: Adapted from R. Dryden, *Before Birth* (London: Heinemann Educational Books, 1978), pp. 13, 14, 15.

FIGURE 6
Recognizing a Face

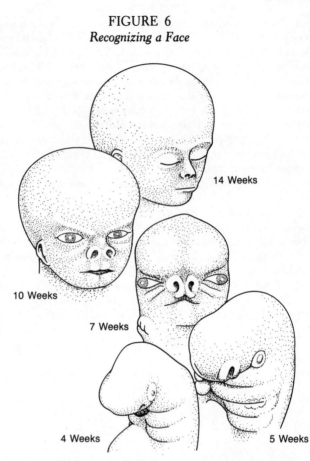

Facial features, so important to recognition both of species and individuals, are unrecognizable as human at seven weeks, transitional at ten weeks, and distinctly recognizable at fourteen weeks.

NOTE: Adapted from R. Dryden, *Before Birth* (London: Heinemann Educational Books, 1978), p. 54.

Appendix

FIGURE 7
Fetal Developmental Course—External Features

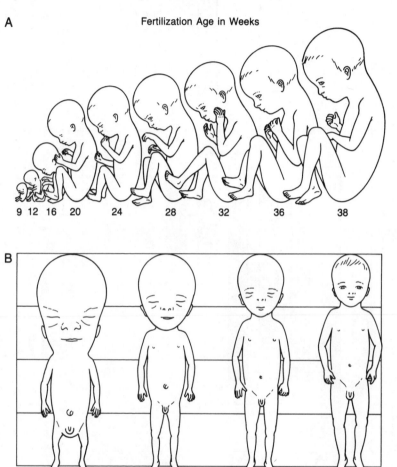

A   Fertilization Age in Weeks

With the major parts and organs already established in the embryonic period, the fetal period is one of growth and maturation in form and function. The growth is illustrated in A, and the maturation of body proportions in B. Between the two parameters, the magnitude of the developmental changes undergone during the fetal period is readily apparent. The increasing functional maturation of the fetus, particularly as displayed in movement, cannot of course be appreciated in static diagrams.

NOTE: Adapted from K. L. Moore, *The Developing Human: Clinically Oriented Embryology*, 3rd ed. (Philadelphia: W. B. Saunders, 1982), pp. 93, 96.

FIGURE 8

Sensitivity to Developmental Abnormality on Exposure to Toxic Agents

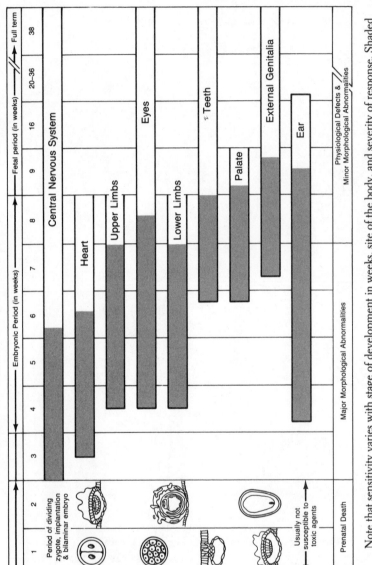

Note that sensitivity varies with stage of development in weeks, site of the body, and severity of response. Shaded bars indicate severe risk, clear bars lesser risk. Overall, the most sensitive period for congenital defects in viable offspring is the third to the eighth week (the period of organogenesis) but the central nervous system, including the brain, remains sensitive throughout the fetal period.

FIGURE 9
Into Circulation

By four weeks the human embryo has established a circuit of flow for a primitive blood that moves through vessels from an exchange point with the mother (early placenta on the left), through vessels leading to embryonic tissues (black), and back through vessels leading to maternal exchange (white). The center of this very early circulatory system is, of course, the heart, the earliest functional organ to develop. Without function of this circulatory system all further development would shortly cease for lack of nutritional resources and accumulation of metabolic wastes.

NOTE: Adapted from R. Dryden, *Before Birth* (London: Heinemann Educational Books, 1978), p. 46.

FIGURE 10
Initiating a Central Nervous System

Not all animals have a central nervous system (CNS), but all complex and active ones do. The most complicated and advanced CNS that we know is our own, and we are only beginning to understand how it works. It is built on principles laid down in our earliest vertebrate ancestors. Probably not unrelated to this fact, CNS construction is initiated right along with establishment of a new developmental individual (see figure 4).

But the structural initiation does not imply establishment of a nervous system in any functional sense. A, B, and C below diagram the genesis of a substrate for neural *function* in a highly oversimplified way.

The first principle of the system is that information from a source (either internal or external) must feed into an information processor (central) and the product must be returned as an output that responds usefully to the input. Useful means, biologically, in the *interest of survival* of the CNS owner.

In abstract information terms a CNS can be diagrammed as follows:

A

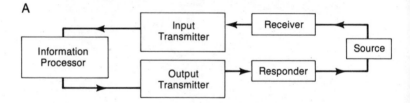

In specific neurological terms, this can be translated into:

B

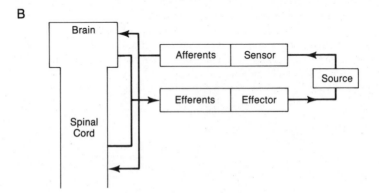

Appendix

C

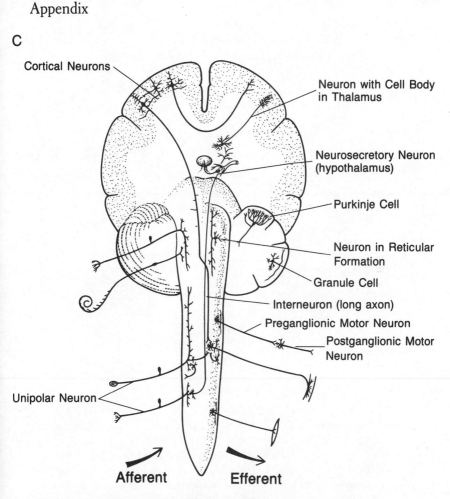

Cortical Neurons

Neuron with Cell Body in Thalamus

Neurosecretory Neuron (hypothalamus)

Purkinje Cell

Neuron in Reticular Formation

Granule Cell

Interneuron (long axon)

Preganglionic Motor Neuron

Postganglionic Motor Neuron

Unipolar Neuron

Afferent   **Efferent**

Above is an idealized presentation of key areas of the brain in relation to the spinal cord and the periphery. On the left are depicted afferent fibers of various neurons that enter the central area from *receptors* at various levels. On the right are depicted efferent fibers from various neurons leaving to terminate on peripheral *effectors*. Interneurons also are shown carrying information within the CNS. All of these neurons transmit information from the periphery to the processing center and back to the periphery in integrated form. They are the *essential* functional elements for all information flow and processing in the CNS.

NOTE: Adapted from C. R. Noback and R. J. Demarest, *The Human Nervous System: Basic Principles of Neurobiology*, 3rd ed. (New York: McGraw-Hill, 1981), p. 50.

FIGURE 11
A Circuit Through the Central Nervous System

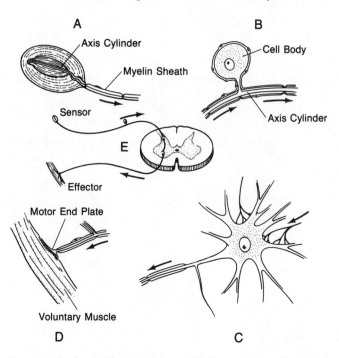

Some details of an informational circuit (E) from sensor to effector through the central nervous system.

A skin sensor for touch (A) contains the tip of a nerve fiber of a bipolar neuron (B) with its cell body in a dorsal spinal ganglion. Its fiber transmits information from the sensor to an interneuron in the spinal cord. In turn, the interneuron transmits information to a large motor neuron (C) with its cell body in the motor area of the spinal cord. A long fiber projected by the motor neuron to the periphery terminates in a motor end plate (D) capable of inciting a contraction of a muscle bundle as a response to the original sensory stimulation.

What is displayed, therefore, is a neuronal circuit (E) that functionally links a receptor to an effector through the spinal cord. The level of information processing in this simple circuit is very low, but it contains the elements that are vastly expanded in more complex circuits (see figure 17).

NOTE: Adapted from C. R. Noback and R. J. Demarest, *The Human Nervous System: Basic Principles of Neurobiology*, 3rd ed. (New York: McGraw-Hill, 1981), p. 51.

Appendix

FIGURE 12
Neurotransmitters at Synaptic Connections

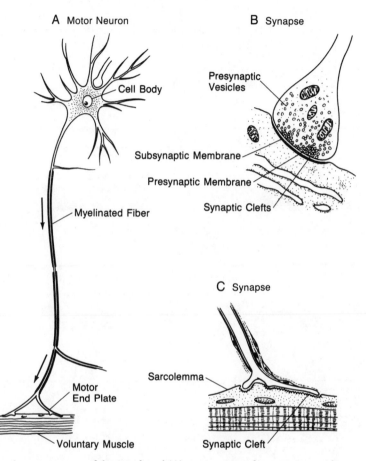

A motor neuron of the spinal cord (A) receives input from synapses with an interneuron on its cell body (B) and transmits the signal along its long process to a motor end plate on muscle (C). Critical steps in the transmission are transport of the information along the fiber and transfer across the synapses at the two functional ends of the neuron on the cell body and at the motor and plate.

NOTE: Adapted from C. R. Noback and R. J. Demarest, *The Human Nervous System: Basic Principles of Neurobiology*, 3rd ed. (New York: McGraw-Hill, 1981), p. 52.

FIGURE 13
Transmission of Signal Along Nerve Fibers and at Synapses

Neural transmission of information occurs in two ways: by propagation of an electrochemical impulse (A) and by chemical movement across a synaptic space (B).

A Electrochemical Impulse Transmission

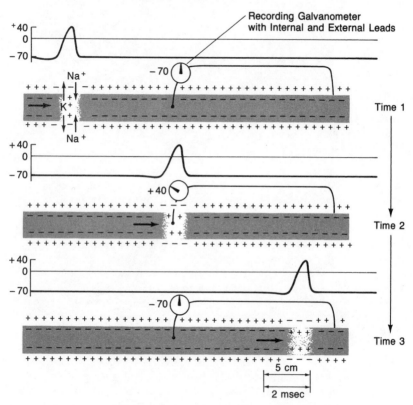

A. A simple model of a nerve fiber at three successive times measured in thousandths of a second (*msec*). The fiber is "wired" to record electrical potential imposed across its external membrane. When the fiber is not carrying a message ("firing"), the potential is negative (−70) because of an excess of negatively charged ions (−) within and positively charged ions (+) on the outside.

When a message moves along the fiber, the external surface membrane becomes momentarily "depolarized"—that is, unequal exchange of charged

Appendix

sodium $(Na+)$ and potassium $(K+)$ ions can occur across the membrane with a resulting shift of potential from -70 to $+40$. The changed potential registers on a recording galvanometer as a "spike," because the charge across the membrane is quickly restored as the impulse passes the recording electrode. Neuronal messages are coded entirely in the number of spikes and the spacing between spikes.

B Synaptic Transmission

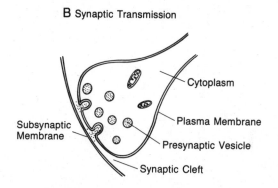

B. When messages reach a synapse they cross it by an entirely different mechanism. The tip of the incoming fiber has numerous secretory presynaptic vesicles containing special neurotransmitter substances. Under the influence of arriving electrochemical impulses, the secretory vesicles discharge their contents into the synaptic cleft, as shown in the diagram. The neurotransmitters diffuse across the cleft and bind specifically to receptors embedded in the subsynaptic membrane of the receptor cell.

In consequence, a new electrochemical disturbance is initiated in the postsynaptic membrane, which propagates again as in A or can trigger some other response such as contraction in a muscle cell. This synaptic mechanism also opens other options essential to higher-level neural integration.

NOTE: Adapted from C. R. Noback and R. J. Demarest, *The Human Nervous System*, 2nd ed. (New York: McGraw-Hill, 1975).

FIGURE 14
Opening the Gates to Interaction and Choice

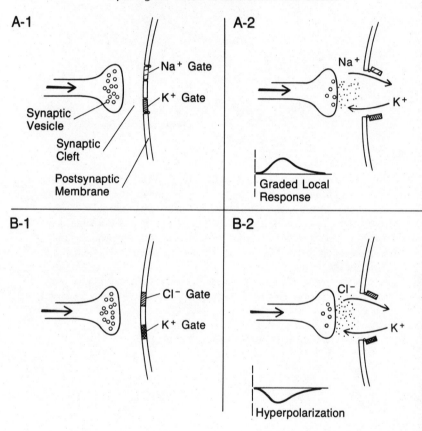

Appendix

Chemical transmission of neural information across synapses opens new possibilities in two major ways: use of alternative chemical neurotransmitters and variations in the response of the postsynaptic (subsynaptic) membrane. The diagram shows how an arriving electrochemical message can have opposite effects (stimulatory versus inhibitory) on the postsynaptic membrane.

Two arriving stimuli can have opposite effects on the same neuron because ion flow is controlled by "gating" of particular ions across the postsynaptic membrane. In A, only sodium ($Na+$) and potassium ($K+$) gates are "opened," and the ion exchange is similar to that described for transmission along a fiber—a depolarization.

But it is also possible to have negative ion transfer as well. In B, gates are opened not only for positive ions but for *negative chloride* ions ($Cl-$) as well. The net effect of movement of oppositely charged ions in both directions is hyperpolarization, with inhibition of transmission as the result. Since a given neuron may receive hundreds of messages, some stimulatory and some inhibitory, its response will be an integral of information from different sources and sensory modalities (touch, taste, vision, hearing, etc.). When one considers the enormous number of neuronal interconnections involved in the human central nervous system, especially the brain, the computational power generated is staggering.

NOTE: Adapted from C. R. Noback and R. J. Demarest, *The Human Nervous System* (New York: McGraw-Hill, 1975), p. 102.

FIGURE 15
Synaptic Integration of Information

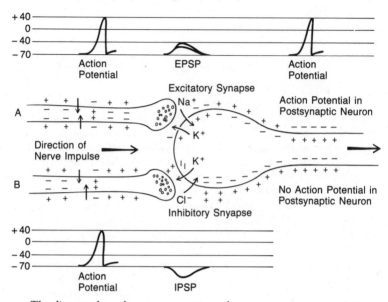

The diagram shows how two synapses on the same neuron, one excitatory and one inhibitory, might interact to affect the outcome in terms of continuing transmission.

The upper electrical tracing shows that an action potential reaching an excitatory synapse (A) produces a small potential change (*EPSP*) in the post-synaptic membrane that is sufficient to initiate a new action potential to be transmitted on. But when a second action potential reaches an inhibitory synapse (B) at the same time, it induces a small change of potential of opposite sign (*IPSP*). The two changes of potential cancel out and no new action potential is initiated.

NOTE: Adapted from C. R. Noback and R. J. Demarest, *The Human Nervous System* (New York: McGraw-Hill, 1975), p. 103.

Appendix

FIGURE 16
Simple Reflex Arcs

Chains of neurons—to, within, and from the central nervous system—provide the base, model, or substrate for simple forms of behavior, which are referred to as reflexive. In turn, the presence, modification, or absence of such behaviors become diagnostic indicators of neurological maturation, normalcy, or damage. The following neurological diagrams illustrate increasing complexity of reflex neural arcs.

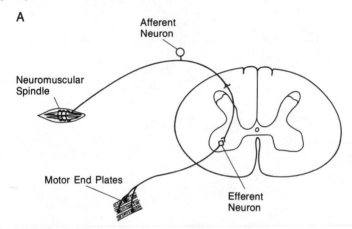

A. A two-neuron reflex arc. An afferent and an efferent neuron synapse in the spinal cord, accounting for contraction of a muscle that is stretched, as in the knee-jerk reflex. Such reactions can help maintain posture without awareness.

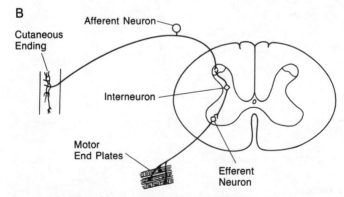

B. Introduction of one or more interneurons within the cord and brain opens an array of more complex interconnections between afferent and efferent. This underlies the simplest form of behavioral reflex, the hot-stove reflex.

NOTE: Adapted from M. L. Barr and J. A. Kiernan, *The Human Nervous System: An Anatomical Viewpoint*, 5th ed. (Philadelphia: J. B. Lippincott, 1988), pp. 78, 80.

FIGURE 17
Complex Afferent Circuitry for Pain

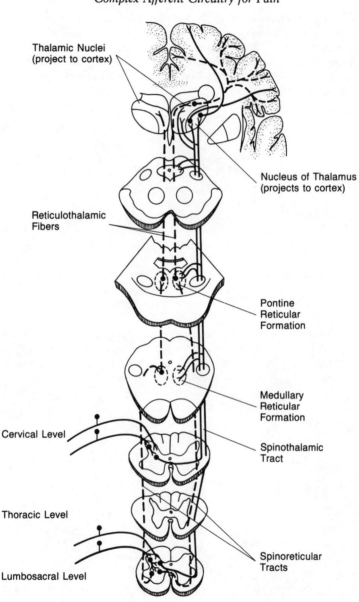

Thalamic Nuclei
(project to cortex)

Nucleus of Thalamus
(projects to cortex)

Reticulothalamic
Fibers

Pontine
Reticular
Formation

Medullary
Reticular
Formation

Cervical Level

Spinothalamic
Tract

Thoracic Level

Spinoreticular
Tracts

Lumbosacral Level

Appendix

This diagram shows the complex (but much simplified!) circuitry underlying the sensation of pain. Inputs from the periphery at several segmental levels (lumbosacral, thoracic, cervical) take place through afferent first neurons that synapse with second interneurons in the spinal cord (as in a simple segmental reflex arc). The interneurons send processes up the spinal cord in several different bundles (tracts) as diagrammed.

One path synapses in the reticular formation of the medulla, another synapses terminally in the thalamus, but with several collateral branches along the way into other areas of the reticular formation. The thalamus, in its turn, sends extensive neuronal processes to synapse with cortical neurons in their highly interconnected network.

The diagram shows only the input (or spinocortical) side of the circuit. The output (or corticospinal) side descends from the cortex to impact on afferent motor neurons at the segmental level of the cord. This, then, is a vastly expanded and complex "reflex arc"—with the brain as the processor between information input and output.

Somehow, in the process, awareness, choice, and purposeful volition emerge. Underlying and essential to these higher functions and behaviors is a neural substrate whose localizations and specific contributions are only dimly understood. But the minimal requirement is for differentiated neurons, synaptic connections, neurotransmitters, and their orderly interactions.

NOTE: Adapted from M. L. Barr and J. A. Kiernan, *The Human Nervous System: An Anatomical Viewpoint*, 5th ed. (Philadelphia: J. B. Lippincott, 1988), p. 283.

FIGURE 18
Tracts in the Spinal Cord

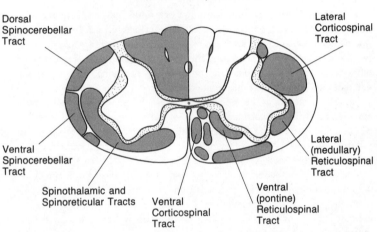

Dorsal
Spinocerebellar
Tract

Lateral
Corticospinal
Tract

Ventral
Spinocerebellar
Tract

Lateral
(medullary)
Reticulospinal
Tract

Spinothalamic and
Spinoreticular Tracts

Ventral
Corticospinal
Tract

Ventral
(pontine)
Reticulospinal
Tract

This diagram gives an indication of the complexity of the packing of inter-neurons of common origin and destination in the white matter of the spinal cord (outside the butterfly-shaped gray matter that contains mostly neuronal cell bodies). Each marked area is a slice across a bundle of fibers. Such a slice or other trauma in a living individual would specifically eliminate particular neurologically based activities. Conversely, in development those activities cannot be present, in the spinal cord or the brain, until the specific neural substrate has matured and is functional.

NOTE: Adapted from M. L. Barr and J. A. Kiernan, *The Human Nervous System: An Anatomical Viewpoint*, 5th ed. (Philadelphia: J. B. Lippincott, 1988), p. 75.

Appendix

FIGURE 19
Genesis of the Human Brain

In this set of diagrams, the ones to the left (3½ to 11 weeks) are at one level of magnification, whereas the ones to the right (13 weeks to newborn) are at a substantially reduced magnification (compare 11 with 13 weeks).

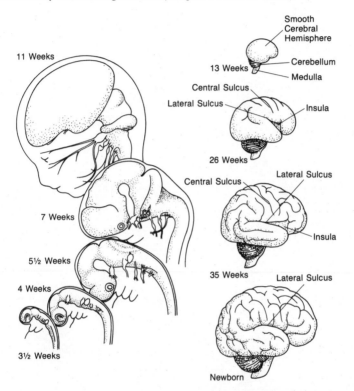

Three things are important to note in this series of stages of human brain development: (1) The central nervous system begins as a continuous tube and remains so even in its complex later stage when the dominance of the cerebral lobes hides virtually all else. (2) The structural dominance of the cerebral lobes is not achieved until the end of the first third of pregnancy (eleven to thirteen weeks). (3) The mass of the cerebral cortex (the outer layer), however, continues its rapid increase during the remainder of the fetal period, as indicated by the surface folding (sulcus) that substantially expands the cortical surface area. This late massive growth of the cortex involves continuing cellular proliferation that must precede full-scale neuronal differentiation and function.

NOTE: Adapted from R. Dryden, *Before Birth* (London: Heinemann Educational Books, 1978), p. 42, and K. L. Moore, *The Developing Human: Clinically Oriented Embryology*, 3rd ed. (Philadelphia: W. B. Saunders, 1982), p. 401.

NOTES

Chapter 1.   The Significance of Status

1. Congregation for the Doctrine of the Faith, *Instruction on Respect for Human Life in Its Origin and the Dignity of Procreation: Replies to Certain Questions of the Day* (Vatican City: Vatican Polyglot Press, 1987).
2. Ibid., p. 5.
3. Ibid.
4. Ibid., p. 9.
5. Ibid., p. 10.
6. Ibid., pp. 12–14.

Chapter 2.   Becoming an Individual

1. D. Tranel and A. R. Dimasio, "Knowledge Without Awareness: An Autonomic Index of Facial Recognition Prosopagnosics," *Science* 228 (1985): 1453–54.

Chapter 4.   Policy for the Unborn: The Preembryo

1. The Ciba Foundation, *Human Embryo Research: Yes or No?* (London and New York: Tavistock Publications, 1986), pp. 149–50.
2. Ibid., pp. 14, 15.
3. Ibid., p. 150 (emphasis added).
4. Congregation for the Doctrine of the Faith, *Instruction on Respect for Human Life in Its Origin and the Dignity of Procreation: Replies to Certain Questions of the Day* (Vatican City: Vatican Polyglot Press, 1987).
5. Roe v. Wade, 410 U.S. 113 (1973); see also Doe v. Bolton, 410 U.S. 113 (1973), and Griswold v. Connecticut, 381 U.S. 479 (1965).
6. J. Robertson, "Ethical and Legal Issues in Cryopreservation of Human Embryos," *Fertility and Sterility* 47, no. 3 (1987), 371–81.
7. See "Research News: Dramatic Results with Brain Grafts," *Science* 237 (1987): 245–47.
8. For a broad presentation of current opinion, see *Hastings Center Report* 17, no. 3 (1987), special supplement.

9. R. Edwards and P. Steptoe, A *Matter of Life* (New York: William Morrow, 1980).

10. See Ciba Foundation, *Human Embryo Research*.

11. Ibid., p. 8.

12. Ibid., pp. 214–15.

13. "The Second Report of the Voluntary Licensing Authority for Human Fertilization and Embryology" (London: Medical Research Council, 1987).

Chapter 5.   Policy for the Unborn: The Embryo

1. P. Hager, "Death Penalty Argument Focuses on Fetal Death," *Los Angeles Times*, May 8, 1987.

2. See J. Schachter, "Infant's Death May Provide a Key Legal Test," *Los Angeles Times*, October 1, 1986.

3. Quoted in ibid.

4. E. Goodman, "When a Child Is Born Ill, Who Is Accountable?" *Los Angeles Times*, October 7, 1986.

5. American Civil Liberties Union of Southern California, "Open Forum," November–December 1986.

6. M. Konen, "Women's Groups Hail Fetal-Neglect Decision," *San Diego Tribune*, February 27, 1987.

7. D. N. James, "Ectogenesis: A Reply to Singer and Wells," *Bioethics* 1, no. 1 (1987): 80–99.

8. Ibid.

9. Ethics Advisory Board, "HEW Support of Research Involving In Vitro Fertilization and Embryo Transfer: Report and Conclusions" (Washington, D.C.: Department of Health, Education, and Welfare, 1979), p. 107.

10. Ibid., p. 111.

11. "Ethical Considerations of the New Reproductive Technologies," *Fertility and Sterility* 46, no. 3, suppl. 1 (1986): 56S, 57S.

12. Ibid., 57S.

13. See the series of reports issued by the National Commission for the Protection of Human Subjects of Biomedical and Behavioral Research (Washington, D.C.: U.S. Government Printing Office, 1975–78).

14. Congregation for the Doctrine of the Faith, *Instruction on Respect for Human Life in Its Origin and the Dignity of Procreation: Replies to Certain Questions of the Day* (Vatican City: Vatican Polyglot Press, 1987), pp. 16, 17, 18.

15. J. Robertson, "Embryo Research," *University of Western Ontario Law Review* 24, no. 1 (1986): 16, 37.

Notes

Chapter 6. Policy for the Unborn: The Fetus

1. B. Siegel, "Why Not Try to Fix What Is Wrong?" *Los Angeles Times*, November 18, 1986.

2. A. C. Rerkin, "Rare Fetus Condition Forces Twins' Parents to Make Tough Choice," *Los Angeles Times*, July 30, 1986.

3. A. R. Jonsen et al., "Critical Issues in Newborn Intensive Care: A Conference Report and Policy Proposal," *Pediatrics* 55, no. 6 (1975): 756–68.

4. Ibid., 757.

5. A. Rostain, "Deciding to Forgo Life-Sustaining Treatment in the Intensive Care Nursery: A Sociologic Account," *Perspectives in Biology and Medicine* 30, no. 1 (1986): 132.

6. L. L. Bailey et al., "Baboon-to-Human Cardiac Xenotransplantation in a Neonate," *Journal of the American Medical Association* 254, no. 23 (December 20, 1985): 3321–29.

7. G. J. Annas, "From Canada with Love: Anencephalic Newborns as Organ Donors?" *Hastings Center Report* 17, no. 6 (1987): 36–38.

8. A. Capron, "Anencephalic Donors: Separate the Dead from the Dying," *Hastings Center Report* 17, no. 1 (1987): 5–8.

9. F. A. Chervenak et al., "When Is Termination of Pregnancy During the Third Trimester Justifiable?" *New England Journal of Medicine* 310 (1984): 501–4.

10. "Hospital Wins Order to Keep Woman Alive," *Los Angeles Times*, July 26, 1986.

11. G. Annas, "Protecting the Liberty of Pregnant Patients," *New England Journal of Medicine* 316, no. 19 (1987): 1213–14.

12. V. E. B. Kolder et al., "Court-Ordered Obstetrical Interventions," *New England Journal of Medicine* 316, no. 19 (1987): 1192–96.

13. Ibid., 1194.

14. 410 U.S. 113 (1973).

15. M. Flower, "Neuromaturation of the Human Fetus," *Journal of Medicine and Philosophy* 10, no. 3 (1985): 237–51.

16. A. L. Gesell, *The Embryology of Behavior: The Beginnings of the Human Mind* (1945; reprint, Westport, Conn.: Greenwood Press, 1971), pp. 108–110.

Chapter 7. Reaching Decisions on Status for the Unborn

1. See the series of reports issued by the National Commission for the Protection of Human Subjects of Biomedical and Behavioral Research (Washington, D.C.: U.S. Government Printing Office, 1975–78).

2. President's Commission for the Study of Ethical Problems in Medicine and Biomedical and Behavioral Research, "Summing Up" (Washington, D.C.: U.S. Government Printing Office, 1983).

3. "Ethical Considerations of the New Reproductive Technologies," *Fertility and Sterility* 46, no. 3, suppl. 1 (1986): 75S.

4. See A. Etzioni, *Genetic Fix* (New York: Harper and Row, 1973), and J. Goodfield, *Playing God* (New York: Random House, 1977).

5. M. Warnock, A *Question of Life* (Oxford: Basil Blackwell, 1985), p. 75.

6. "The Second Report of the Voluntary Licensing Authority for Human Fertilization and Embryology" (London: Medical Research Council, 1987).

7. See *Fertility and Sterility* 46, no. 3, suppl. 1 (1986): appendix D.

8. A. Rostain, "Deciding to Forgo Life-Sustaining Treatment in the Intensive Care Nursery: A Sociologic Account," *Perspectives in Biology and Medicine* 30, no. 1 (1986): 117–35.

9. Ibid., 122.

10. Ibid., 124–25.

11. Ibid., 132.

Chapter 8. Epilogue

1. B. R. Finney and E. M. Jones, eds., *Interstellar Migration and the Human Experience* (Berkeley: University of California Press, 1985).

INDEX

Abnormalities, prematurity and, 114, 115; *see also* Defects

Abortion, *vii*, *viii*; due to anencephalic fetus, 119; "burnout" among health professionals assisting in second trimester, 36; debate over, 12, 17–19, 20; for defects in embryo or fetus, 89; without destroying embryo, 142; external embryos from, 92–93; external embryos from, alternate uses of, 100; fetal viability and, 109; generating policy for, 38; issues of status of unborn beyond, *xi–xii*; "morning-after" approach as, 63; pain of unborn in, 51; rape and, 15; *The Silent Scream*, 33, 36; spontaneous, 77; tension and dissention over, 6–7; therapeutic, 102–3; in U.S., number of, 87

Adoption of preembryo, 73, 151

Afferent circuitry for pain, 188–89

Afferent (sensory) fibers, 43, 48, 179

Alcoholism and birth defects, 91

Ambiguity of status of unborn, *viii*

American Civil Liberties Union (ACLU), Reproductive Freedom Project of, 91

American Fertility Society, 152; Ethics Committee of, *xiv*, 62, 94–95

Amniocentesis, 110

Amniotic cavity, 169

Anencephaly, 75, 89, 118–19

Annas, George, 120

Arousal, 56

Attention, 56

Autoimmune deficiency, 102

Autonomic division of nervous system, 51

Awareness: inner, status and, 144–46; interactive, 144; *see also* Sentience

Axes, body, 27, 83, 171

Axial structural organization of nervous system, 47–48

Basic motor repertoire, 128–31

Behavior: as foundation for fetal status, 127–28; movement as first component of, 33, 37, 85, 155; neural integration and, 41–42; psychic individuality and, 34; as special aspect of function, 30–32

Behavioral individuality, 22, 29, 30–32, 33, 37, 38, 40, 107, 125

Beneficence, principle of, 119

Biomedical Ethics Board, 147

Biomedical science, dramatic advance of, *vii*

Biotechnology, space technology and, 162

Birth as status marker, 121–23

Birth defects, *see* Defects

Blastocyst, 26–27, 47, 60, 168, 169

Blood circulation in embryo, 177

Index

73; ethical and legal issues in, 64–65; space travel and, 158–62

Culture, fusion of perspectives derived from science and technology with, *xiii–xiv*

Culture (medium): of preembryonic human cells, 68–71, 79–80; production of eggs from ovaries in, 101

Custodianship for preembryos, 78

Death, anencephaly and definition of, 118–19

Decision process, nature of, *x*; *see also* Status of unborn

Declaration of Independence, 66

Defects: abortion due to, 89; alcoholism and, 91; anencephaly, 75, 89, 118–19; developmental abnormalities, 103, 176; diagnostic access to embryo to assess, 88–89, 102–3; frequency and nature of, 89–90; loss rate of zygotes due to genetic, 77; prenatal diagnosis of, 88–89, 102–3, 110–11; prevention of embryonic, 89–92; toxic substances causing, 92, 103–4, 176

Designed status, 153–56

Development, human: coping with enigma of, 135–36; different aspects of, *ix*; of embryo, external features of, 172–73; of fetus, external features of, 175; IVF and accessibility of earliest stages of, *xi–xii*; major phases of, *xii–xiii*; and sensitivity to exposure to toxic agents, 92, 103–4, 176; as translevel phenomenon, 37–39

Developmental abnormalities, 103, 176; *see also* Defects

Developmental individuality, 22, 26–28, 36, 38, 47, 62, 94

Diagnostic access to embryo, 88–89, 102–3

Dignity of human life, objective of preserving, 17

Discomfort, *see* Pain

Dispositional authority for preembryos, assignment of, 74–75

DNA, 24, 26, 37, 65, 137; recombinant, 148–49

Down's syndrome, 89, 102

Drug testing in nonhuman species, reliability of, 104

Drug use during pregnancy, 90–92

Ectogenesis, 93, 94, 97, 100

Effectors, 179

Efferent (motor) fibers, 48, 179

Egg, 168; fertilization of, 8, 15–16, 24, 81, 82, 136, 168; fertilized, inherent potential of, 133–34

Egogenesis, 123–31

Elderly people, senile dementia in, 141

Embryo: augmented value in, 139–43; behavioral characteristics of, 140; development of, 83–84, 85, 172–73; formation during implantation, 82; heart and circulation, development of, 28–29, 177; individuality of, 140, 155; intermediate states between adult complex organisms and, 108; movement of, 50; precursor of, 27; as "prequick," 84; primary embryonic organization, 27; relationship between mother and, 86–87, 139–40; research on, 94, 95–97; scientific concept of, 59; therapeutic transplantation from, 100–101; transition from preembryo to, 96; transition to fetus from, 30, 85, 99

Index

Index

Index

Individuality (*continued*)
nificance of movement to, 32–33, 37, 40; social, 23, 35–36, 37, 41

Individuality, neural substrate of, 31–32, 40–57, 178–91; becoming truly neural, 48–49; behavior and neural integration, 41–42; circuitry, as essential substrate, 49–50; circuitry, complexities of, 45–46; circuitry, sources of, 46–48; minimal inner experience and, 56–57; minimally adequate, 35; neurons and information flow, 42; pain in the unborn, 51–55; recognition of inner experience and, 51–54; spinal reflex arc as prototype of, 43–44; spontaneity versus reflex and, 50–51; synaptic switches, 44–45, 181–83, 186

Infant, 109; social interaction of, 144

Infanticide, 75

Infertility, male, 16

Informational aspect of pain, 52, 53

Information flow, neurons and, 42

Informed consent, 18, 19, 95

Inhibitory input, 43

Inner awareness, status and, 144–46

Inner experience, 54–55; fetal status and, 135; minimal, 56–57; recognition of, 51–54; role of brain stem in, 54

Intensive care nurseries, 113, 144; life decisions in, 116, 152–53

Interactive awareness, 144; *see also* Social interaction

Interneurons, 43–46, 48, 53, 179, 187–90

Intersegmental neurons, 46

Intrauterine device, 9

In vitro fertilization, *vii–viii*, 72; accessibility to earliest stages of human development through, *xi–xii*; British centers for, 81; cryopreservation and, 10, 11–12; in-

dependence of preimplantation phase, 8, 9, 10, 59; preembryos awaiting transfer to maternal uterus in, 66; research on preembryos and, 78–79, 94, 151–52; status of preembryo in, 9, 73; success rate of, 79; surplus eggs obtained in, use of, 101; tension and dissension over, 7; transfer time for preembryos, 93–94

Keene, Barry, 91–92

Kolder, V. E. B., 120

Lesch-Nyhan syndrome, 89

Life cycle, 66–67

Life-support technology, 113, 145, 146; viability and, 126; withdrawal of, 115

McLaren, Anne, 61–62

Malignancy in human preembryonic cells, study of, 68

Mammalian heritage, 7, 108

March of Dimes, 104

Marsupials, fetal period in, 108

Melbourne, Australia: controversy over surgical introduction of sperm into egg, 15–16; cryopreservation case in, 11, 12

Mendeloff, John, *xiv*

Minimal inner experience, 56–57

Mitosis, 168

Morality, public policy and, 14

"Morning-after" pill, 9, 63

Morphogenetic (form-generating) activities, 98

Morula, 168

Mother: difficulty of separating status

Index

icine and Biomedical and Behavioral Research, 147
Prevention of embryonic deficiencies, 89–92
Primary embryonic organization, 27
Primitive streak, 27–28, 62, 83, 169, 171
Privacy rights, reproductive, 87, 109
Progression of states, *xii*; stepwise assignment of status during, *xiii*
Pro-life movement, *see* Right-to-life movement
Pronucleus, 168
Proprioceptive reflexes, 50–51
Prosopagnosia, 34
Psychic individuality, 22, 33–35, 37, 38, 40, 50; in fetal period, 107; fetal status and, 122–23; uncertainty over time of onset of, 34–35; *see also* Self-genesis
Public policy, *see* Policy for unborn

Quickening, 37, 99, 128

Rape, abortion and, 15
Receptors, 179
Recognition: of humanness, empathy and, 98–99, 134–35; of inner experience, 51–54; social individuality and, 35–36; *see also* Social individuality
Recombinant DNA, 148–49
Recombinant DNA Advisory Committee (RAC), 149–50
Reflex arc: simple, 187; spinal, 43–44
Reflex movement: proprioceptive, 50–51; spontaneity versus, 31, 32–33, 50–51
Reproduction, uneasiness over power

of science to intervene and modify, 17–18
Reproductive Freedom Project of American Civil Liberties Union, 91
Reproductive privacy, 87, 109
Research, 150–52; on animal species, reliability of, 104, 112; on embryos, 94, 95–97; fetal, 7, 38, 112; on human subjects, 18, 19; on preembryos, 74–75, 78–81, 82, 94–95; on preembryos, areas in need of, 80; on preembryos, in Great Britain, 80–81; on preembryos, *in vitro* fertilization and, 78–79, 94, 151–52; recombinant DNA controversy, 149
Respiratory distress, 113
Reticular formation, 51, 54, 55, 128–29, 189
Rights: human, 6; of mother, 87, 109; women's, 7
Right-to-life movement, 12; definition of unborn status, 15; position on fetal status, 116; on preserving dignity of human life, 17; simplification of beginning of life, 23–24
RNA, 26
Robertson, John, 64–65, 96
Roe v. Wade, *vii*, 64, 126
Rostain, Anthony, 116, 152–53
Royal College of Obstetricians and Gynaecologists, 81, 152
Rudiment, 47

Science: Congregation for the Doctrine of the Faith on, 12–15; fusion of perspectives derived from culture and technology with, *xiii*–*xiv*; power to intervene and modify reproduction, uneasiness over, 17–18; special role and contribution

205

Index

Science (*continued*)
of, *xi*; *see also* Research; Technology

Self, definition of, 123

Self-awareness, 124; psychic individuality and, 33–35

Self-genesis, 123–31

Self-identity, 122–23; *see also* Psychic individuality

Senile dementia, 141

Sentience: cortical development and, 129; fetal, 124–25, 127–28, 135, 154–56; as minimal inner experience, 56; movement and, 33, 127–28; neural maturation and, 124; viability and, 130; *see also* Psychic individuality

Separation of church and state, 17

Siblings from frozen preembryos, age disparity of, 11–12

Sickle-cell anemia, 102

Silent Scream, The (videotape), 33, 36

Singleness, origin of, 26–28; *see also* Developmental individuality

Skeletal axis, appearance and maturation of major, 29–30

Skin grafting, 100

Social individuality, 23, 35–36, 37, 41

Social interaction: empathy and, 135; of fetus versus infant, 144

Social status: behavioral individuality and, 125; changes in, 154

Somites, 171

Space technology, biotechnology and, 162

Space travel, frozen preembryos and, 158–62; colonization and, 159–62

Sperm, 168

Spinal cord: rudiment of, 47–48; tracts of, 46, 53, 189, 190

Spinal reflex arc, 43–44

Spinothalamic tract, 53, 189

Spontaneous abortion, 77

Spontaneous versus reflexive movement, 31, 32–33, 50–51

Startle response, 32

State and church, separation of, 17

Status: changes in course of life, 6; in democratic society, 6

Status of unborn, 3–20; achieving and implementing, 146–48; ambiguity of, *viii*; based on relative versus absolute criteria, 15, 16; birth as status marker, 121–23; Congregation for the Doctrine of the Faith on, 12–15; consensual objectives on, 16–19; considering individuality in relation to, 36–39; cryopreservation and, 10, 11–12; designed, 153–56; difficulty of separating status of mother from, 7–8; emergent policy pattern, 150–52; inner awareness and, 144–46; as issue, reasons for, 5; making decisions about life, difficulties of, 152–53; model of recombinant DNA and, 148–49; political feasibility of definition of, 19–20; progression of states, *xii*, *xiii*; public, 13, 14–15; reaching decisions on, 132–56; reaching decisions on, complex problems regarding, 135–46; reasons for reevaluating, 133–35; right-to-life definition of, 15; for zygote, 5–6; *see also* Embryo, status of; Fetus, status of; Preembryo, status of

Superovulation, 10–11

Surgery, fetal, 38, 41

Surrogate mother, 151

Synapses, 44–45, 48; chemical transmission of neural information across, 184–85; among cortical neurons, 55; excitatory and inhibitory, 186; frequency, basic motor repertoire and, 31; integration of

Index

information, 186; neural transmission at, 182–83; neurotransmitters, 44, 45, 54, 181, 182

Syngamy, 15–16, 24

Technology: Congregation for the Doctrine of the Faith on, 12–15; consensual guidelines on advancement and application of, 18–19; fusion of perspectives derived from culture and science with, *xiii–xiv*; life-support, 113, 126, 145, 146; recombinant DNA, 148–49; space, biotechnology and, 162

Teratogens, sensitivity to, 86, 92, 103–4, 176

Teratoma, 61

Testing, *see* Research

Thalamus, 53, 54, 55, 129, 189

Thalidomide tragedy, 103

Therapeutic abortion, 102–3

Therapeutic transplantation: of cultured preembryonic cells, 68–69, 74–75; from embryos, 100–101; using organs and tissues from marginally living fetuses, 118–19; *see also* Research

Therapy for unborn: advances in, 111–12; fetal surgery, 38, 41; putting mother at risk, 110–11

Toxic external agents, exposure to, 86, 92, 103–4, 176

Tracts in spinal cord, 46, 53, 189, 190

Transplantation, *see* Therapeutic transplantation

Trophoblast, 60–61, 82, 169

Trusteeship for preembryo, 139

Ultrasound imaging, 32, 88, 110; embryonic movement and, 50; of independent behavior, 37; recording of early fetal movements, 33; staging human development with, 99

Unborn: covert existence of, 7; as negative term, *viii–ix*

Unique and unitary, individual as, 22

U.S. Department of Agriculture, 150

United States Supreme Court, *Roe v. Wade* decision, *vii*, 64, 126

University of Texas School of Law, 64

Vatican Instruction on Respect for Human Life, 63–64, 96

Ventral horn, 31

Viability, fetal, 64, 85; concept of, 109; as foundation for fetal status, 125–27; life-support technology and, 126; rise of, 125; selected time of, 145–46; sentience and, 130; statistical nature of, 126–27

Voluntary Licensing Authority, 81

Warnock Committee, 152

Women: expansion of occupational roles and exposure of, 104; mammalian heritage shaping role and lives of, 7; *see also* Mother

Women's rights, 7

Yolk sac, 169

Zygote, 14, 24, 27, 76, 77, 168; early changes undergone by, 59; losses during natural development, 77; status of, 5–6; *see also* Preembryo